4th International Conference on Technology and Electrical Engineering (CITIE 2021)

Journal of Physics: Conference Series Volume 2135

Online
11 – 12 November 2021

ISBN: 978-1-7138-5106-6
ISSN: 1742-6588

Printed from e-media with permission by:

Curran Associates, Inc.
57 Morehouse Lane
Red Hook, NY 12571

Some format issues inherent in the e-media version may also appear in this print version.

Institute of Physics
Dirac House, Temple Back
Bristol BS1 6BE UK

Phone: 44 1 17 929 7481
Fax: 44 1 17 920 0979

techtracking@iop.org

Additional copies of this publication are available from:

Curran Associates, Inc.
57 Morehouse Lane
Red Hook, NY 12571 USA
Phone: 845-758-0400
Fax: 845-758-2633
Email: curran@proceedings.com
Web: www.proceedings.com

TABLE OF CONTENTS

4th International Conference on Technology and Electrical Engineering (CITIE 2021)

Preface

The curricular project of Technology in Electricity and Electrical Engineering by cycles of the Universidad Distrital Francisco José de Caldas (Colombia) celebrates every two years its international conference. Unlike previous editions, during 2021 the Universidad Distrital has its facilities closed, and with restricted access, as a preventive measure to avoid the spread of the SARS-CoV-2 virus. Probably in two years, for CITIE 2023, the conference will return to its usual physical space in Aduanilla de Paiba, our flagship library. However, for this fourth version of CITIE, we opted for a distance event, organized entirely by videoconference, an experiment that, although not a feature mastered by our community, was a success.

The fourth International Conference on Technology and Electrical Engineering (CITIE 2021) was held on November 11-12, 2021. This event was attended by researchers from the electrical area with papers mainly focused on electronic instrumentation, control, power systems, and electromagnetic compatibility. From the talks and presentations, it was evident that our research community remains active and produces cutting-edge research despite the conditions of restriction and social unrest typical of our political and socio-economic circumstances, strongly affected by the COVID-19 pandemic.

A total of 11 papers were submitted to CITIE 2021, of which 10 were approved by the jurors (conference committee members and international reviewers) in a double-blind evaluation process. The conference was held entirely online over two days, which included the test session, three 50-minute keynote speeches, and three author presentation sessions. The research papers were separated into three thematic sessions, Instrumentation and Control, Power Systems, and Electronic Compatibility. These papers were each given 20 minutes of presentation time, including five minutes for questions. The entire conference was broadcast live on the event's YouTube channel, registering an average of 600 daily visits.

I would like to take this opportunity to acknowledge the hard work of the entire CITIE 2021 committee. Their advice and support were fundamental to preparing for the conference. I would also like to thank the students of the Electrical Technology program who supported the videoconference platform and the live transmission. Likewise, a special thanks to the speakers and authors for considering CITIE 2021 as a means of dissemination of their researchers, and to the invited researchers, whose talks and discussions allowed to leave on the carpet countless ideas for further discussion and development of impact research.

The three previous versions of CITIE allowed the development of activities such as business shows, workshops, and a degree of interaction that was not achieved in CITIE 2021. The face-to-face conferences allow a level of direct communication with authors and researchers that is difficult to replicate on a distance platform. Even so, CITIE continues to maintain its spirit of dissemination of research and innovation, as well as its function of identifying and promoting areas of research in the field of electrical energy. We are confident that this version of the event has provided a great experience for all attendees, increased interest in the research fields discussed, as well as renewed expectations for the CITIE 2023 meeting.

Best regards,

Prof. Fredy Martínez, Ph.D

CITIE 2021

Universidad Distrital Francisco José de Caldas

COMMITTEES

Program Chairs

Prof. Fredy Martínez, Ph.D (Universidad Distrital, Colombia)

Prof. César Chacón, Ph.D (Universidad Distrital, Colombia)

Technical Committee

Journal of Physics: Conference Series **2135** (2021) 011001 doi:10.1088/1742-6596/2135/1/011001

Prof. César Hernández, Ph.D (Universidad Distrital, Colombia)

Prof. Henry Ibañez, MSc (Universidad Distrital, Colombia)

Prof. Luis Noguera, MSc (Universidad Distrital, Colombia)

Prof. Yaqueline Garzón, MSc (Universidad Distrital, Colombia)

Prof. Edwar Jacinto, MSc (Universidad Distrital, Colombia)

Prof. Fernando Martínez, MSc (Universidad Distrital, Colombia)

Prof. Edwar Jacinto, MSc (Universidad Distrital, Colombia)

Prof. Diego Acero, MSc (Universidad Pedagógica Nacional, Colombia)

Peer Review Statement

All papers published in this volume have been reviewed through processes administered by the Editors. Reviews were conducted by expert referees to the professional and scientific standards expected of a proceedings journal published by IOP Publishing.

Type of peer review: Double Anonymous

Conference submission management system: Morressier

Number of submissions received: 11

Number of submissions sent for review: 11

Number of submissions accepted: 10

Acceptance Rate (Submissions Accepted / Submissions Received × 100): 90.9 %

Average number of reviews per paper: 2.45

Total number of reviewers involved: 7

Contact person for queries:

Name: Fredy Martínez

Affiliation: Universidad Distrital Francisco José de Caldas

Email: fhmartinezs@udistrital.edu.co

Journal of Physics: Conference Series **2135** (2021) 012001 doi:10.1088/1742-6596/2135/1/012001

Autonomous identification of high-contact surfaces from convolutional neural networks

Angie Alonso[1], Andres Peña[2], and Fredy Martínez[3]

Universidad Distrital Francisco José de Caldas, Facultad Tecnológica, Cra. 7 No.40B-53, Bogotá D.C., Colombia

E-mail: [1]analonsol@correo.udistrital.edu.co, [2]anfpenar@correo.udistrital.edu.co, [3]fhmartinezs@udistrital.edu.co

Abstract. The rapid spread of the SARS-CoV-2 virus has highlighted many social interaction problems that favor the spread of disease, particularly airborne spread, which can be addressed by adjusting existing systems. Of particular interest are places where large numbers of people interact, as they become a focus for the spread of these diseases. This paper proposes and evaluates an autonomous identification scheme for certain surfaces considered high risk due to their continuous handling. These high-contact surfaces can be identified by an autonomous system to apply specific cleaning tasks to them. We evaluate three convolutional models from a proprietary dataset with a total of 2000 images ranging from wall switches to water dispensers. The objective is to identify the ideal architecture for the system. The ResNet (Residual Neural Network), DenseNet (Dense Convolutional Network), and NASNet (Neural Architecture Search Network) models were selected due to their high performance reported in the literature. The models are evaluated with specialized metrics in non-binary classification problems, and the best scheme is selected for prototype development.

1. Introduction

Contamination of high-contact surfaces has now become a focus of special attention to minimize the spread of viruses, including SARS-CoV-2 and others such as enteric viruses [1]. These viruses can remain on surfaces for several weeks, or even months, and act as sources of contagion and potential spread [2]. The surfaces at risk are classified as porous and non-porous, the latter referring for example to stainless steel and plastics, among others, and are surfaces on which it has been proven that viruses can remain for a longer time [3]. The permanence of viruses on these surfaces depends on the frequency with which they are cleaned, a strategy that has been implemented for the benefit of public health is the disinfection of surfaces more routinely, and increased hand hygiene practices to reduce viral transmission in any space with high interaction of people [4].

One of the main sectors affected by the current pandemic has been the tourism sector [5]. Closures, quarantines, and mobility controls have considerably reduced the economic movement of this sector [6]. For example, the World Tourism Organization (WTO) estimates a worldwide drop in tourism of between 20% and 30% [7, 8]. For this reason, the hotel sector is in the task with implementing strategies to encourage visitors to return to their facilities safely. In this context, the care and maintenance of high-touch surfaces are of vital importance, both to provide safety and to prevent them from becoming a source of contagion [9].

The adopted strategy of this research contemplates the training of a series of classification models through the use of artificial neural networks (ANN) and their possible application in the processing of images related to this type of surfaces. Artificial neural networks date back to 1943 with the first neural model presented by McCulloch and Walter Pitts, and from which the starting point for the development of current deep neural models emerged [10, 11]. The application area we will focus on is image processing, a process that includes several levels. First of all, there is a pre-processing where the information of an image is adapted, and better profiling of the parameters is to be analyzed and identified [12]. For example, brightness and contrast, which can hide or highlight important features in the images. Secondly, segmentation, where the image is partitioned into several regions to better represent the information in detail [13]. The third is the detection and classification of the objects found in the image, and finally the analysis of the image where we obtain the most information of what the image shows [14].

Our objective is to find a high-performance convolutional model that can be integrated into an artificial system with digital cameras, to guide maintenance processes in high contact surfaces [15]. In this sense, it is pertinent that the data obtained by the neural network correspond to the characteristics of the problem, which involves the real-time operation and high capacity of identification in real human environments. Neural networks are capable of identifying specific common parameters in the input data and allow obtaining a closed system that produces output without performing complex mathematical calculations. In this sense, it is convenient to use them in a low-cost embedded system that does not require high processing and/or memory requirements, and that allows fast-updating according to supervised training. Similar schemes have been used to support the fight against SARS-CoV-2, most of them in the development of embedded systems that allow the diagnosis of the disease, but they can be used in other contexts such as the one proposed here [16–18].

Our problem is focused on spaces related to tourist environments, in general, indoor spaces characteristic of hotels and similar. To solve the problem we will describe the methodology to be used both to create the database and to train the different convolutional models. These models will be evaluated using performance metrics to identify the most suitable architecture for use in an embedded prototype [19–21]. In the final part, the results of each model are discussed, and schemes for model improvement and further research are proposed.

2. Problem statement

The tourism sector has grown exponentially in recent years and has become a reference in the world economy. However, nowadays, and with the crisis that arose due to the SARS-CoV-2 pandemic, this sector was faced with a generalized crisis, and its vulnerability to effects that directly affect it became evident. Studies have shown that high-touch surfaces such as door handles, light switches, or laptop computers can spread the virus more rapidly to other surfaces, and people, if they are not properly sanitized. For this reason, it is important to implement strategies to make economic activity compatible with the need to contain the virus and avoid outbreaks.

The objective of this research is to go a step beyond traditional cleaning services and to be able to offer a disinfection service that allows, hand in hand with technology and research on neural networks, its use in real environments with a high turnover of people. The aim is to develop an image-based categorization model capable of autonomously identifying critical elements for a cleaning process in hotels and equivalent spaces. This strategy seeks to regain the confidence of tourists when they return to these vacation destinations, providing them with assurance about the cleaning policies. The methodology proposed to address the problem is shown in the flow chart in Figure 1.

The recognition system must be able to learn the parameters of interest from a proprietary dataset developed by the research group for this purpose. This means that the model must

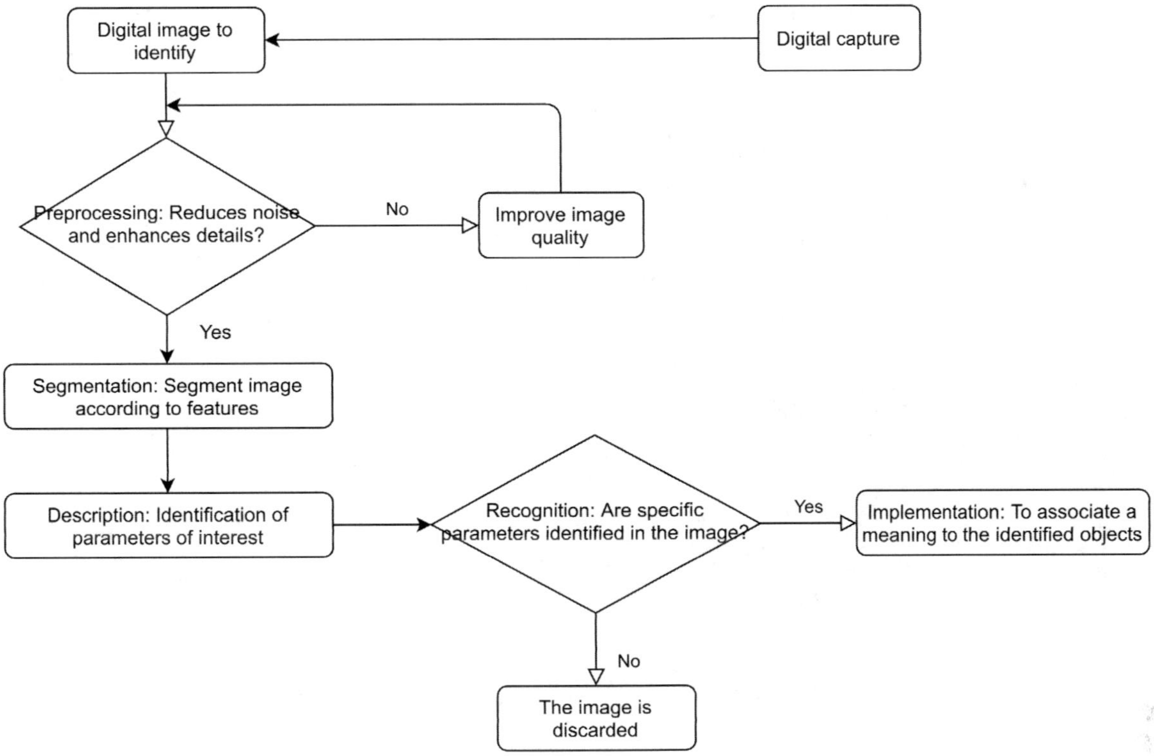

Figure 1. Structure of the system.

be able to identify specific parameters in the images regardless of their location, additional elements in the image, size, or orientation. The evaluation of the classification model is performed using the metrics Accuracy, Precision, Recall, F1-score, and confusion matrix. In addition, this evaluation must consider both average performance and performance by category.

3. Methods

By design, convolutional neural networks correspond to receptive fields of neurons in the visual cortex of a biological brain. Consequently, it is assumed that they can identify parameters in images in a similar way as humans do. Therefore, a convolutional model must be trained with a large set of images containing the features that make them different, in the same way that humans learn to identify elements throughout their lives.

Among the different existing convolutional models, we have selected three for their previous high performance on similar problems. We selected the ResNet (Residual Neural Network), DenseNet (Dense Convolutional Network), and NASNet (Neural Architecture Search Network) architectures to evaluate their performance with our dataset, and select the most suitable one to implement our autonomous identification system. The ResNet type networks are characterized by implementing blocks with information jumps forward in the network, which has been observed to increase the capacity of the system. Following the same idea, the DenseNet architecture performs these same jumps forward but in a denser way than the ResNet model, and instead of adding the information it concatenates it, achieving an important reduction in parameters. One of the main improvements of DenseNet is that it manages to shorten the connections between layers near the input and output, thus increasing the density of the network. Finally, the NASNet architecture is characterized by blocks optimized for certain databases, which are then generalized to the network and other similar classification problems. The three selected models have the possibility of using reduced-size models that would eventually facilitate their implementation on an embedded system.

Our dataset is composed of 20 categories, which include elements typically found in hotels such as wall switches, wall sockets, handrails, door handles, window, and drawer handles, among others. Each category had a total of 100 images, some downloaded from public repositories, and others captured in the laboratory. Figure 2 shows an example of one of the categories of our dataset. All images were subjected to preprocessing consisting of channel adjustment, contrast enhancement, and scaling to a size of 256×256 pixels. Image preprocessing was performed with OpenCV 4.1.2 This dataset was then randomly separated into two groups, 80% of the images were used for model training, and the remaining 20% for final model validation.

Figure 2. Sample of the dataset.

The training of the three models was performed in the same way. In all three cases, we started with a set of random weights adjusted with the SGD (Stochastic Gradient Descent) function as optimizer, and Categorical Cross Entropy as loss function. The training was performed over 80 epochs, and in each epoch, the accuracy and MSE (Mean Squared Error) values for both the training data and the validation data were checked. This process was used to make fine adjustments to the training, and to improve its performance in each case. All code was developed in Python 3.7 with support for Keras 2.6.0 and Tensorflow 2.6.0.

4. Results

The first results were obtained during the training process of the three models (Figures 3, 4 and 5). In all three cases, the models reached their highest performance point at a maximum of 40 epochs. Although in the following epochs a slight improvement was observed in all three cases, the benefit achieved does not justify the computational cost. The three models reached high Accuracy values for the training data (above 90%), however, they had different behaviors for the validation data. While the validation data were unknown for all three models, this tracking throughout training allowed the behavior to be adjusted to benefit the generalizability of the models. This worked very well with the ResNet and DenseNet models, but not with the NASNet model. The validation data for the ResNet and DenseNet models reached 30%, a value that, although low, exceeded the 10% achieved by the NASNet model. According to these behaviors, overfitting of the NASNet model is observed.

We calculated three metrics on the models using the validation data: Precision, Recall, and F1-score. The first one determines how well the model classifies items in a category concerning

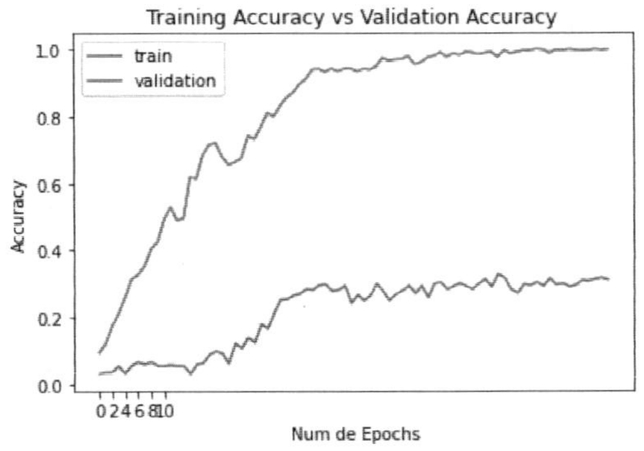

Figure 3. Accuracy for training and validation data. ResNet model.

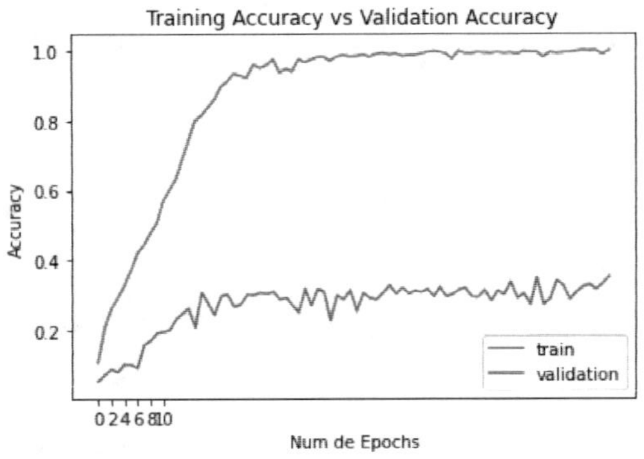

Figure 4. Accuracy for training and validation data. DenseNet model.

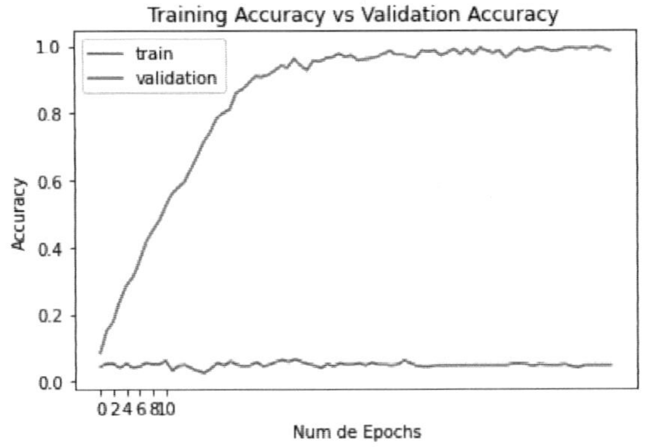

Figure 5. Accuracy for training and validation data. NASNet model.

the total, the second one is more specific with the classified elements that actually belong to the category, and the third one is a weighted average of the first two metrics. Tables 1, 2, and 3 show the results of these metrics. As observed from the accuracy, the ResNet and DenseNet models perform much better with new data than the NASNet model. In fact, the NASNet model classified all images in category 2, so it got right those that actually belonged to this category, but missed all others. Consequently, the NASNet model does not meet the minimum performance requirements. The ResNet and DenseNet models perform much better with validation data and are very similar in each of the categories. This result validates them as promising in system development and raises the question of the quality of the initial database.

Although the ResNet and DenseNet models achieve some generalization capability in the presence of unknown data, their performance remains low compared to the expected performance. Throughout the performance, the three models manage to reduce their error and reach accuracy above 90% in all cases, but it is difficult for them to go above 30% with the validation data. According to these results, it can be inferred that there is no overfitting of the ResNet and DenseNet models, but that it is necessary to increase the complexity of the training database to provide the models with more information when working with unknown datasets.

Journal of Physics: Conference Series **2135** (2021) 012001 doi:10.1088/1742-6596/2135/1/012001

Table 1. Metrics by category and average. ResNet model.

	Precision	Recall	F1-score	Support
0	0.18	0.13	0.15	20
1	0.28	0.21	0.24	20
2	0.36	0.39	0.37	20
3	0.26	0.32	0.29	20
4	0.26	0.29	0.28	20
5	0.22	0.32	0.26	20
6	0.17	0.15	0.16	20
7	0.40	0.43	0.41	20
8	0.21	0.18	0.19	20
9	0.12	0.20	0.15	20
10	0.38	0.59	0.47	20
11	0.26	0.35	0.30	20
12	0.44	0.30	0.36	20
13	0.78	0.52	0.62	20
14	0.44	0.22	0.29	20
15	0.60	0.55	0.57	20
16	0.27	0.21	0.24	20
17	0.50	0.50	0.50	20
18	0.21	0.25	0.23	20
19	0.16	0.19	0.18	20
Weighted avg.	0.33	0.31	0.31	400

The last metric used on these models was the confusion matrix. Again, the validation data was used to evaluate the behavior of the models. To facilitate their analysis, a heat code was incorporated in which light colors indicate a higher quantity of elements, and dark colors a lower quantity; consequently, a well-defined light-colored diagonal should be observed in the best models. The results are shown in Figures 6, 7, and 8. As observed with the previous metrics, the best performances were achieved with the ResNet and DenseNet models, with DenseNet significantly better than ResNet, while it was verified that the NASNet model misclassifies most of the images in category 2. In general terms, the DenseNet model has an acceptable performance with validation data, and reduced size compared to ResNet, which makes it suitable for the development of the embedded identification system. Even so, the low capacity of the models to correctly infer categories for unknown data raises the need to evaluate the structure of the database used, probably by considerably increasing its size to increase the training data.

5. Conclusions

This paper presents the development of an autonomous model for the classification of high contact surfaces as a strategy for the application of specific maintenance and cleaning strategies in hotels and accommodation spaces with a high turnover of people, to ensure hygiene and health safety against possible disease contagion. The model is based on training from images containing characteristic elements of these environments such as electrical switches or doorknobs. According to the required performance, and the ability of the model to operate in real-time on an embedded device, we proposed the evaluation of three convolutional architectures that have reported high performance in this type of task. The ResNet, DenseNet, and NASNet topologies were selected and trained with their own dataset of 2000 images under the same conditions. The

Table 2. Metrics by category and average. DenseNet model.

	Precision	Recall	F1-score	Support
0	0.09	0.09	0.09	20
1	0.36	0.33	0.35	20
2	0.41	0.30	0.35	20
3	0.19	0.20	0.19	20
4	0.32	0.26	0.29	20
5	0.20	0.26	0.23	20
6	0.29	0.19	0.23	20
7	0.50	0.36	0.42	20
8	0.20	0.18	0.19	20
9	0.18	0.40	0.25	20
10	0.27	0.53	0.36	20
11	0.48	0.42	0.45	20
12	0.36	0.15	0.21	20
13	0.67	0.74	0.70	20
14	0.46	0.38	0.41	20
15	0.68	0.68	0.68	20
16	0.24	0.37	0.29	20
17	0.45	0.68	0.55	20
18	0.40	0.40	0.40	20
19	0.35	0.27	0.30	20
Weighted avg.	0.36	0.35	0.35	400

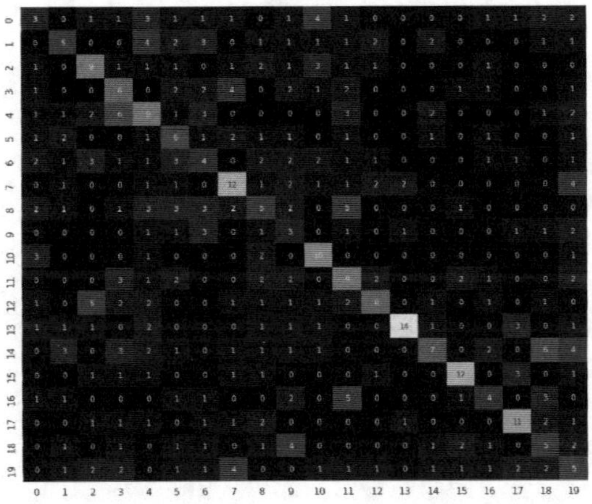

Figure 6. Confusion matrix. ResNet model.

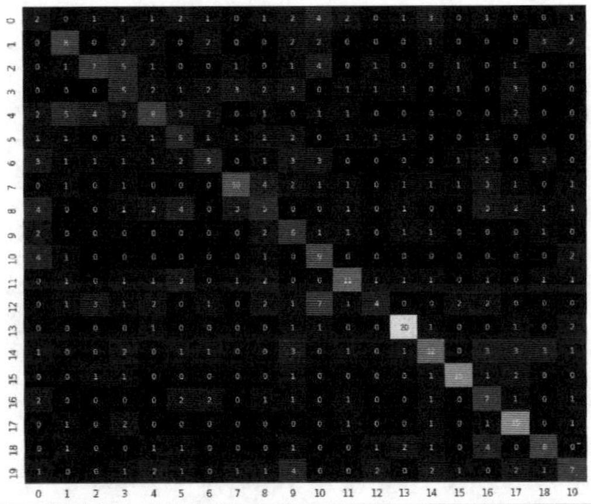

Figure 7. Confusion matrix. DenseNet model.

three models were evaluated with the same metrics using a small set of the dataset not used in the training. The results show that the NASNet model is unable to generalize its behavior to unknown data, an activity better performed by the ResNet and DenseNet models. According to the results, the DenseNet model is selected as the architecture due to its high performance and smaller size compared to the ResNet model. Additional results include the need to optimize the DenseNet model in an embedded system with a larger and more complex database.

Table 3. Metrics by category and average. NASNet model.

	Precision	Recall	F1-score	Support
0	0.00	0.00	0.00	20
1	0.00	0.00	0.00	20
2	0.05	1.00	0.09	20
3	0.00	0.00	0.00	20
4	0.00	0.00	0.00	20
5	0.00	0.00	0.00	20
6	0.00	0.00	0.00	20
7	0.00	0.00	0.00	20
8	0.00	0.00	0.00	20
9	0.00	0.00	0.00	20
10	0.00	0.00	0.00	20
11	0.00	0.00	0.00	20
12	0.00	0.00	0.00	20
13	0.00	0.00	0.00	20
14	0.00	0.00	0.00	20
15	0.00	0.00	0.00	20
16	0.00	0.00	0.00	20
17	0.00	0.00	0.00	20
18	0.00	0.00	0.00	20
19	0.00	0.00	0.00	20
Weighted avg.	0.00	0.05	0.00	400

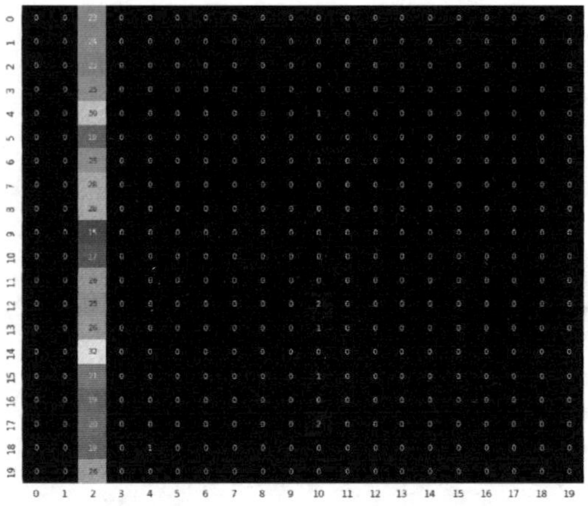

Figure 8. Confusion matrix. NASNet model.

Acknowledgments
The authors wish to acknowledge and thank the Technological Faculty of the Universidad Distrital Francisco José de Caldas for the support and funding during the course of this research work.

References
[1] Mendes M, Oliveira A, Pires O, Branca F, Beirao M, Santa A, Carvalho A and Alves J 2021 *Acta Med Port* **34** 1–7 ISSN 0870-399X
[2] Brass A, Shoubridge A, Crotty M, Morawska L, Bell S, Qiao M, Woodman R, Whitehead C, Inacio M, Miller

C, Corlis M, Larby N, Elms L, Sims S, Taylor S, Flynn E and Papanicolas L 2021 *BMC Infect Dis* **21** 1–7 ISSN 1471-2334
[3] Cuevas E, Girón I, Falcó I, Pérez A, Díaz A, Aznar R, Randazzo W and Sánchez G 2021 *Environmental Research* **203** 111831 ISSN 0013-9351
[4] Elbasi E, Topcu A and Mathew S 2021 *Electronics* **10** 1–18 ISSN 2079-9292
[5] Brouder P 2020 *Tourism Geographies* **22** 484–490 ISSN 1470-1340
[6] Martínez L, Barreto M and Wanomen L 2015 *Tekhnê* **11** 5–8 ISSN 1692-8407
[7] Díaz C 2021 *Turismo y Sociedad* **29** 183–199 ISSN 0120-7555
[8] Montiel H, Martínez F and Gómez H 2020 *International Journal of Engineering Research and Technology* **13** 1934–1939 ISSN 0974-3154
[9] Ibn-Mohammed T, Mustapha K, Godsell J, Adamu Z, Babatunde K, Akintade D, Acquaye A and Fujii H 2021 *Resources, Conservation and Recycling* **164** 105169 ISSN 0921-3449
[10] Varela E and Campbells E 2011 *Investigación y desarrollo en TIC* **2** 18–27 ISSN 0121-3261
[11] Khan A, Sohail A, Zahoora U and Qureshi A 2020 *Artificial Intelligence Review* **53** 5455–5516 ISSN 0269-2821
[12] Ramírez J and Chacón M 2011 *Revista de Ingeniería Eléctrica, Electrónica y Computación* **9** 7–16 ISSN 1870-9532
[13] Mennel L, Symonowicz J, Wachter S and Polyushkin D 2020 *Nature* **579** 62–66 ISSN 0028-0836
[14] Ibtehaz N and Rahman M 2020 *Neural Networks* **121** 74–87 ISSN 0893-6080
[15] Zhang Y, Liu Y, Sun P, Yan H, Zhao X and Zhang L 2020 *Information Fusion* **54** 99–118 ISSN 1566-2535
[16] Tripathi A, Jain A, Mishra K, Pandey A and Vashist P 2020 *Revue d'Intelligence Artificielle* **34** 673–682 ISSN 0992-499X
[17] Martínez F, Martínez F and Jacinto E 2020 *Int J Adv Sci Eng Inform Technol* **10** 662–667 ISSN 2088-5334
[18] Liu P, Tsai Y, Chen P, Tsai H, Hsu L, Wang C, Lee N, Huang M, Wu Y, Ko W, Yang Y, Chiang J and Shen M 2020 *Journal of Medical Internet Research* **22** 19878 ISSN 1438-8871
[19] Díaz J and Pérez M 2004 *Tekhnê* **2** 9–17 ISSN 1692-8407
[20] Galvis J and Madrid J 2016 *Tekhnê* **13** 43–48 ISSN 1692-8407
[21] Zhang J, Jin X, Sun J, Wang J and Sangaiah A 2020 *Multimedia Tools and Applications* **79** 15095–15115 ISSN 1380-7501

Parallel control model for navigation tasks on service robots

Holman Montiel[1], Fernando Martínez[2], and Fredy Martínez[3]
Universidad Distrital Francisco José de Caldas, Facultad Tecnológica, Cra. 7 No.40B-53, Bogotá D.C., Colombia
E-mail: [1]hmontiela@udistrital.edu.co, [2]fmartinezs@udistrital.edu.co, [3]fhmartinezs@udistrital.edu.co

Abstract. Autonomous mobility remains an open research problem in robotics. This is a complex problem that has its characteristics according to the type of task and environment intended for the robot's activity. Service robotics has in this sense problems that have not been solved satisfactorily. These robots must interact with human beings in environments designed for human beings, which implies that one of the basic sensors for structuring motion control and navigation schemes are those that replicate the human optical sense. In their normal activity, robots are expected to interpret visual information in the environment while following a certain motion policy that allows them to move from one point to another in the environment, consistent with their tasks. A good optical sensing system can be structured around digital cameras, with which it can apply visual identification routines of both the trajectory and its environment. This research proposes a parallel control scheme (with two loops) for the definition of movements of a service robot from images. On the one hand, there is a control loop based on a visual memory strategy using a convolutional neural network. This system contemplates a deep learning model that is trained from images of the environment containing characteristic elements of the navigation environment (various types of obstacles and different cases of free trajectories with and without navigation path). To this first loop is connected in parallel a second loop in charge of defining the specific distances to the obstacles using a stereo vision system. The objective of this parallel loop is to quickly identify the obstacle points in front of the robot from the images using a bacterial interaction model. These two loops form an information feedback motion control framework that quickly analyzes the environment and defines motion strategies from digital images, achieving real-time control driven by visual information. Among the advantages of our scheme are the low processing and memory costs in the robot, and the no need to modify the environment to facilitate the navigation of the robot. The performance of the system is validated by simulation and laboratory experiments.

1. Introduction

Events such as social isolation due to the SARS-CoV-2 virus, coupled with the tendency of society's long working hours to leave children, the elderly, and people with medical needs at home alone for long periods, bring to the table the growing need for automated systems to provide specialized care, training, and monitoring services in the home. Service robots are one solution to this problem, and the great advances made to date suggest that they will have a strong impact on our society. However, these robots still have many unresolved problems, such as those related to their autonomous navigation in unfamiliar environments, and which have been built for humans [1].

The human being can navigate comfortably in this type of environment relying mainly on the sense of sight. This is the main sensor for navigating in both familiar (home, workplace, etc.) and unfamiliar (shopping mall, bank office, etc.) environments. The signals picked up by the eye are then processed at different levels, in some cases to reason a response, in others to reflexively avoid danger [2]. Accordingly, it is natural to think that a service robot, which will navigate in the same environments, should have similar capabilities to enable it to navigate efficiently in these environments [3]. This principle should provide it with a system with functional performance similar to that of a human, at least in terms of autonomous movement.

Previous experiences constitute a large database that human beings use to recognize new events and elements, comparing them with those already known, and thus defining their interaction with them, it is in this way that previous experiences define their way of navigating in an environment [4]. This can be seen in simple everyday examples, if a person is in a hallway, he can walk down the hallway until he reaches, for example, a door and enters the room, and perform this task while simultaneously dodging obstacles such as pets, boxes or other obstacles in the way. It is also easy for him to identify his destination, and therefore recognize when he reaches it, which is natural when he first enters the bathroom of a hotel or an unfamiliar house. The task is performed by identifying different visual features such as specific shapes, colors, or even combinations of these. Given the similarity of the problem, a robot would also succeed in performing these tasks if it uses a similar control scheme, this scheme is known as visual memory (VM) [5].

The way to implement this structure is similar to the biological model, it is to consolidate over two stages. During the first stage, a training process is carried out using images with characteristic elements of the environment [6]. Some of the images correspond to obstacles from different perspectives, while others correspond to free paths through which the robot can navigate. All these images correspond to labels related to the expected response of the robot. In this way, the learning process corresponds to the creation of an autonomous visual classifier. This process is carried out off-line; the second stage corresponds to an on-line system of propagation and classification of the images by the robot. This stage is much faster and seeks to provide additional support to the first stage.

The functionality of such a parallel control scheme is its high functional performance and reduced computational cost. This scheme is very efficient in structured and static environments, and even with a certain level of dynamics. It has the great advantage of not requiring a complete mapping of the environment, reducing the problem to reactive behavior through Euclidean point reconstruction, which considerably reduces processing requirements, while also reducing response times [7]. Therefore, it can be implemented in parallel on the robot with other interaction functions without a reduction in performance.

This scheme is very consistent with the biological model. Autonomous navigation strategies based on image feedback turn out to have a high performance in real applications. The images contain all the information needed to navigate safely, even in dynamic environments, and do not require prior knowledge of the environment [8]. According to the parameters to be extracted from the images, the processing tends to be fast, allowing real-time operation with very small intervals between processed images (short response time) [9]. These features are ideal for a robot that has to respond simultaneously to several criteria.

One problem, however, of these schemes is in the discontinuity between images generated by the movements of the robot. It is complex to identify the same parameters between two images after the robot's movement. One solution is the use of specific references in the environment [10], another option is to generate autonomous responses from each image without the need to link the information between different images. For the latter, it is necessary to extract more information from the environment from each image, but the advantage is that the environment is not modified to guarantee the robot's navigation [11].

In this paper we propose an autonomous navigation scheme for service robots in structured indoor environments along predefined paths in the environment, using images through the VM strategy. The learning of the geometric constraints of the environment is done on a convolutional neural network. To avoid the problem of the discontinuity between images caused by the robot movement, we propose a second control loop in parallel with the VM loop supported by stereoscopic vision and geometric location of relevant points of the obstacles in front of the robot. It is well known that a stereoscopic optical system (with two cameras side by side) outperforms the single-camera [12] system due to its depth perception capability, a key element of our parallel control loop [10, 13–15].

This second control loop will provide specific information related to the distance and position of the objects in the environment, again taking as a reference the human optical system, that is, implementing a visual sensor system formed by two simultaneous images of the same obstacle from the perspective of the robot, but displaced a small distance to define information related to the depth [15]. With the assistance of the system geometry, it is possible to determine precisely the distance of the robot to the obstacle (or a point on the obstacle) from the obstacle distances in the two images, taking into account the focal distance of the cameras [16].

The basic idea is to represent the spatial geometry (x, y, and z) of the environment and the 3D projections of objects in this space (z-dependent distance functions), from the two 2D perspectives (stereoscopic vision) that simulate the human optical system. In most of the schemes reported in the literature, this process is computationally expensive [17] and needs additionally a very good calibration of the cameras, which in real-time applications on an embedded system becomes almost impossible to achieve [18]. Another important element in the strategy is the quality of the lighting in the 2D images, as this has a strong impact on the 3D model [19]. The algorithm is a back-projection from the 2D images (considering the angular distance and the distance between cameras) to obtain the 3D localization of a point in the space [15, 20]. Many applications, such as ours, do not require the complete three-dimensional reconstruction of the environment; it is enough to identify some points of the obstacles, which considerably reduces the computational cost [21]. Again concerning the biological model, we have similar behavior in the human brain when it processes information from the eyes, since it only processes part of the optical information (in which the focus and attention are located). The information collected is then contrasted with the VM loop to define specific parameters of the environment to allow safe navigation [22, 23].

In terms of strategies for estimating distances to obstacles in stereoscopic vision systems, there are two trends [14, 24]. In the first one, there is an active strategy, since the sensors send signals to the obstacles (using a laser or visible light), and their reflection is detected and analyzed [15]. The great advantage of this strategy is the precision and versatility to establish distances, much higher than the human one. However, the system as such is expensive, slow, and complex to implement real applications. Also, a laser system can establish distances to an obstacle point, but sweeps are required to establish the general topology of the obstacle [25]. The second trend is supported by passive methods, which estimate the location of the obstacle by processing the information recorded in two stereoscopic images [21, 26]. Traditionally these images are processed digitally and allow a certain degree of independence concerning the type of cameras required. The classic problems of occlusion and overlap of elements in the images are still solved by parallel strategies [27].

Our research is oriented to the development of service robots, in particular in applications related to the care and monitoring of people (children and elderly). In this sense, the performance of the robot must be characterized by safe interaction with humans, robust in the development of tasks, and high performance at low cost [19, 25]. In this work the last two criteria are of special interest, the robot must perform the navigation along the path identifying obstacles and free paths as it moves, and the processing of the algorithms must be performed on the robot in real-time. In this sense, the strategy we propose uses a double loop control scheme to guarantee robustness, and the algorithms used allow real-time execution on embedded hardware. Other advantages of our strategy are that there is no need for complete planning and its high performance in any environment [28].

2. Problem statement

Considering the robot environment as one characterized by indoor spaces spatially limited by walls, obstacles (static and dynamic) and doors that can give access (or not, depending on its configuration) to other regions of the environment, our problem is focused on the definition of motion for an autonomous robot in that environment from the analysis in real-time of environmental images captured from the perspective of the robot, with characteristics of low resource consumption and high reliability in the development of the task.

The robot must move along a path drawn in the environment. Geometrically we define as $\mathscr{W}$ the workspace of the robot, which is an open subset of $\mathbb{R}^3$ and is much larger than the robot, so the robot can move without problems in the free space $E \in \mathscr{W}$. This free space corresponds to the free interior of $\mathscr{W}$, which is defined as $\mathscr{W} - \mathcal{O}$, where $\mathcal{O}$ corresponds to the set of all obstacles in $\mathscr{W}$ that represent volumes not accessible to the robot. These obstacles are finite in size and quantity and are visually identifiable by the robot. The walls of the environment that limit and shape $\mathscr{W}$ correspond to the boundaries of $\mathscr{W}$ and are denoted as $\partial\mathscr{W}$.

To identify the path, obstacles, and $\partial\mathscr{W}$, the robot has two digital cameras that form a stereoscopic vision system. This system is located in $r(t) \in \mathbb{R}^3$ and has $R(t) \in SO(3)$ orientation, where $SO(3)$ denotes the special orthogonal group of dimension three with respect to a global frame of reference for every instant $t \geq 0$.

The estimate of the location of these elements in the environment is made concerning a relative reference frame, whose origin is defined in relation to the digital cameras installed in the robot (i.e. concerning the robot's field of vision, Figure 1). These cameras are labeled as Left Camera (L_C) and Right Camera (R_C). According to the robot dimensions (the robot used as platform is the ARMOS TurtleBot developed by the research group), the central position of L_C and R_C to the relative reference frame are (-0.14, 0, 0) and (0.14, 0, 0). The horizontal distance between the centres of the two cameras is $b = 0.14 + 0.14 = 0.28$ m.

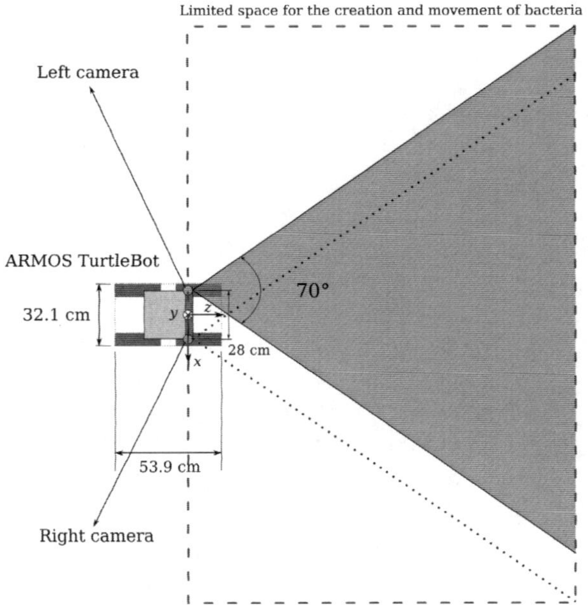

Figure 1. Reference frame relative to the robot used to estimate the location of points of the obstacles. The figure corresponds to the elevation view and contains the geometrical relations of the experiments.

The navigation of the robot requires the robot to recognize the path to follow, and at the same time to recognize the obstacles to avoid along the way. To achieve the robot's movement a set of movement policies is defined, which determine the behavior of the two robot motors from the visual identification of the path and the obstacles in the environment. The use of two control loops for visual identification is proposed. A first loop capable of identifying characteristics of the environment in front of the robot from the training of known conditions (images with the path, free spaces for navigation and obstructed spaces with different obstacles), and a second loop that determines concrete points of the obstacles (estimated distance to the obstacle) through geometric analysis of the stereoscopic images.

Each of the obstacles in $\mathcal{O}$ is indexed by $i \in \mathscr{H} = \{1, 2, 3, ..., n\}$. These are identified by the first control loop, but their exact position is determined by the second control loop. For the classification of images by training, we propose the use of a convolutional neural network, specifically a Dense Convolutional Network (DenseNet). This network is characterized by a dense connection, fewer parameters, and high accuracy compared to other convolutional networks such as ResNet (Residual Neural Network). The design of its structure is based on the concept that the accuracy of the network increases if it contains shorter connections between the layers near the input and the layers near the output, consequently the network concatenates each output of the previous layers forward (Figure 2).

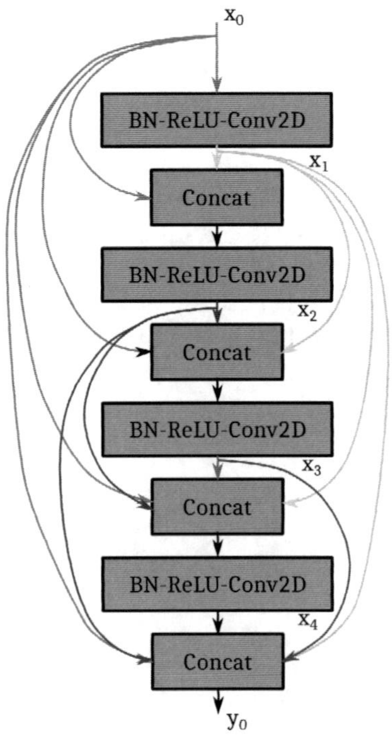

Figure 2. Densely Connected Convolutional Network (DenseNet) architecture.

The position of the obstacle $\mathcal{O}_i$ concerning the global reference frame is given by (equation 1):

$$x_i(t) = R(t)\,\boldsymbol{p}_i(t) + r(t) \tag{1}$$

Where $\boldsymbol{p}_i(t)$ is the position of the obstacle to the robot. For an obstacle in the three-dimensional system, this information corresponds to the set of points that belong to the obstacle. The determination of all the points of the obstacle carries a high computational cost, and its calculation is not necessary for our algorithm. Instead, we propose to identify a small group of points of the obstacle that allows us to estimate with precision the distance of the obstacle to the robot, and therefore to define the movement policy more adapted to follow the established path of navigation.

To search and accelerate the convergence to these points we use a search algorithm based on a simplified model of bacterial interaction developed by the research group. In our search algorithm, we defined a bacterial population (m bacteria in total, without considering reproduction and death processes), which move in the state space conformed by the three-dimensional space $\mathscr{W}$ according to local interaction rules. For the search, the bacteria are initially located in $\mathscr{W}$, and the movement update is done according to the information provided by the two parallel images of the cameras. If a bacterium is identified on an obstacle, then we can estimate the direct distance of the robot to the obstacle, and consequently, plan the movement of the robot in the environment (Figure 3).

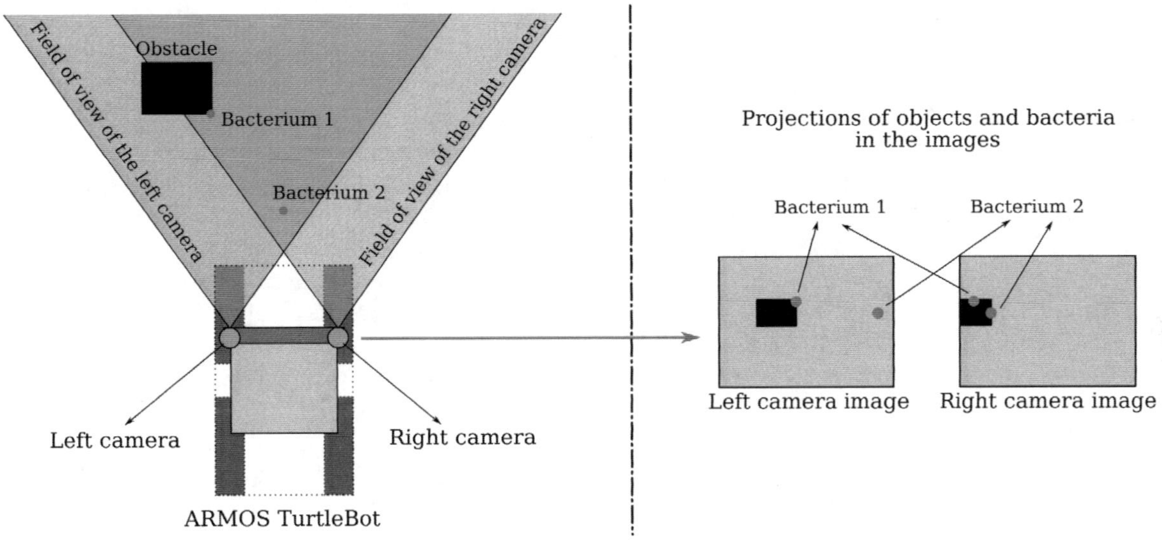

Figure 3. Densely Connected Convolutional Network (DenseNet) architecture.

The obstacles correspond to locations in $\mathscr{W}$ of high interest for bacteria (equivalent to areas with food concentration). Thus, when a bacterium finds an obstacle, it continues to explore the neighboring region to identify other bacteria in it. A region is more attractive to other bacteria if there are several bacteria in it, this process is reinforced by the activation of Quorum Sensing (QS), which means that if the amount of bacteria in a region of $\mathscr{W}$ exceeds a quorum threshold T, then the performance of the region is increased. In our previous research, we have found that QS reduces convergence time by at least 30%.

Population density is evaluated using a norm to establish the distance between bacteria. For example, a bacterium is defined as $V = (p)$, where p is a point in three-dimensional space, and the distance between bacteria V_i and V_j is defined as $d(V_i, V_j)$. As norm, the Euclidean, Lebesgue or Sobolev distance can be used.

The algorithm to solve the three-dimensional location of the bacteria from the two-dimensional images evaluates the similarity of the neighboring pixels to the bacteria in the two camera images according to the following equation (equation 2):

$$F = \frac{|\nabla (M_L)| \cdot |\nabla (M_R)|}{\sum_{color} \sum_{(i,j) \in N} \left[L_{(x_{L+i}, y_{L+j})} - R_{(x_{R+i}, y_{R+j})} \right]^2} + f\,(QS) \tag{2}$$

Where (x_L, y_L) and (x_R, y_R) correspond to the coordinates of the left and right projections of the bacterium (Figure 3), L corresponds to the grayscale value of the pixel addressed in both the left and right image, N is the neighborhood around the evaluated bacterium, and $D(M_R)$ is Sobel gradient norm over the left and right projections, which is intended to penalize uniform regions.

3. Methods

Our autonomous navigation system proposal is designed for applications on small service robots in indoor environments with sensorial limitations (no access to GPS and limited WiFi communication) and processing limitations due to the need for real-time operation and the development of multiple tasks simultaneously (in addition to the desire to reduce its total cost). These features have forced us to develop minimalist, high-performance solutions. In particular, the problem addressed corresponds to the development of a navigation scheme that involves large displacements of the robot autonomously in known and structured environments.

As a solution, we propose a motion control scheme based on images by recognizing and tracking landmarks in the environment located in such a way that together they specify a navigation path. As tracking patterns, we have experimented with different types of landmarks on walls and floor of the environment, but in this paper, we document the highest performance scheme, a colored line drawn on the floor of the environment. This pattern is identified by the control unit, as well as the constraints of the environment involving walls, doors, and obstacles. The control scheme is based on feedback of information generated from the geometric characteristics of the landmarks and the geometric restrictions of the environment, in both cases resulting from a stereoscopic vision system (epipolar geometry) consisting of two digital cameras.

The proposed scheme consists of two control loops that take as input parameters the same information, the images of the two cameras of the stereoscopic system (Figure 4). However, neither of the two loops requires decomposition of explicit pose parameters, a feature that is generally used in similar investigations, and which considerably increases the processing demands on the control unit.

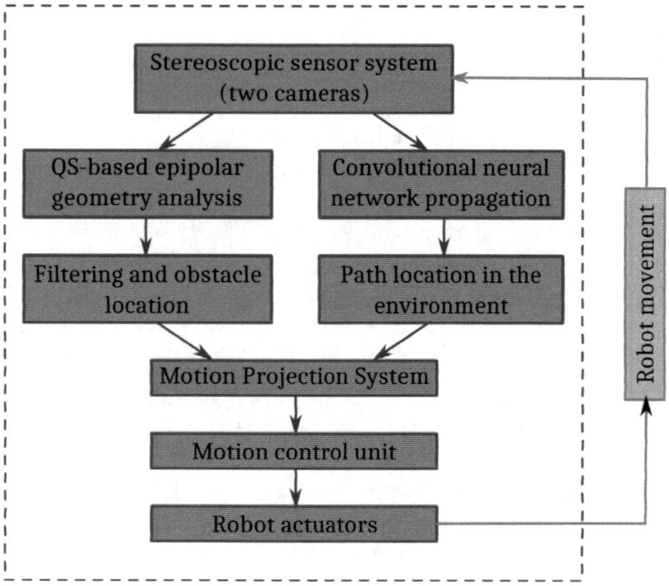

Figure 4. Proposed navigation scheme for autonomous robots based on stereoscopic vision and convolutional network.

The first loop corresponds to a VM system trained with images captured from the actual navigation environment in our laboratory. The images were separated into four categories:

(i) Front of the robot without obstacles and with navigation landmark.

(ii) Front of the robot without obstacles and without navigation landmark.
(iii) Front of the robot with obstacles and navigation landmark.
(iv) Front of the robot with obstacles and without navigation landmark.

Each category was identified with the label corresponding to the number and was made up of 200 images (800 images in total, of which 80% were used for training, i.e. 640 images). The VM system model was built with a DenseNet type convolutional neural network. We chose the DenseNet network because it is more efficient in terms of parameters compared to other topologies such as ResNet, maintaining equivalent classification performance.

The motion control unit analyses the motion projection derived from the two control loops and determines the direction and distance to be covered. It does this according to control policies in correspondence with the results of the two control loops. For example, if the two loops define that there are no obstacles in front of the robot, that the navigation landmark is in front of it, and the distance to obstacles is large, then the advance distance is adjusted and the robot is ordered to advance. If on the other hand there are obstacles or it does not detect the landmark, it must rotate on its axis and look for more information. Before advancing the robot must identify the landmarks near its position and locate an obstacle-free path.

The motion speed is constant in the robot, and its activation on the motors is adapted to the identified path and distances. The images are taken simultaneously and under controlled conditions from the robot, i.e. the robot is kept still and without vibrations (this is verified from the control unit and guaranteed by using an accelerometer). In this way, the quality of the images is guaranteed and discontinuities are avoided.

The second control loop calculates the depth of points on obstacles using stereoscopic vision. In short, stereoscopic vision produces a geometry called epipolar geometry, characterized by two two-dimensional projections corresponding to a single three-dimensional environment. Recently we have developed a work related to the 3D reconstruction of obstacles in front of a robot, for the fast and precise calculation of distances from the robot to the obstacle through stereographic optical sensors. This system allows us to predict with high reliability the structure of an obstacle from two parallel images that provide depth information. We build on previous work that provides algorithms for the reconstruction of stereoscopic data from 2D digital camera images and put those same algorithms into a scalable, low-cost computational framework. Our approach is more efficient than most, because it only identifies a limited set of points of the obstacles, offering a partial reconstruction of the 3D vision in real-time.

4. Results

The VM-based control loop was implemented using a convolutional neural network. This solution was chosen due to its high capacity to differentiate characteristics of the environment from images and its easy propagation on the robot. The training was done off-line using real images from the laboratory.

In particular, we used a DenseNet network of 121 layers deep, these layers were organized in a $5 + (6 + 12 + 24 + 16) \times 2 = 121$ structure, where 5 is (conv, pooling) + 3 transition layers + classification layer. Multiplication by 2 is done because each dense block has two layers (1×1 conv and 3×3 conv). The number of nodes in the input layer is determined by the shape of the images. We scaled all input images to 255×255 pixels and processed them in RGB, so the shape of the input layer was $255 \times 255 \times 3$. The number of nodes in the final output layer is the number of possible class labels, in this case, the output layer had four nodes. With this structure, the neural network required 6,957,956 trainable parameters and 83,648 non-trainable parameters.

As loss function, we use categorical cross-entropy, as optimizer we use stochastic gradient descent, and as metrics to evaluate the performance of the neural model we calculate the accuracy, recall, f1-score, and support (Figure 5 and table 1). Also to evaluate the model performance, we generate the confusion matrix and ROC curve of the trained network (Figure 6, 7, and 8). The training was carried out during 10 epochs, and it was verified that there was no over-adjustment.

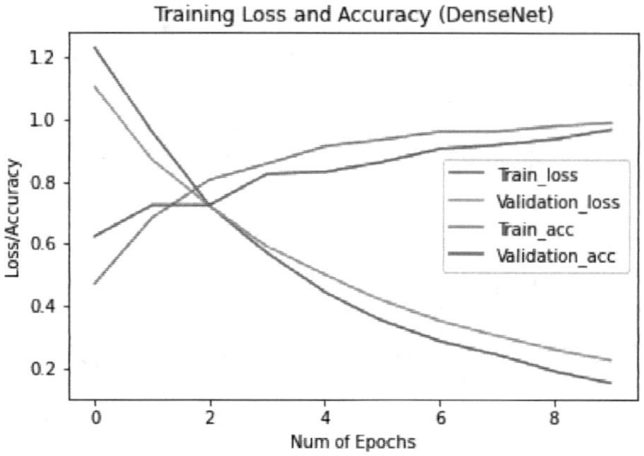

Figure 5. Model behavior based on training and validation data.

Table 1. Model metrics.

	Precision	Recall	F1-score	Support
0	0.98	0.92	0.95	48
1	0.92	0.98	0.95	45
2	1.00	1.00	1.00	32
3	1.00	1.00	1.00	35
Macro avg.	0.97	0.97	0.97	160
Weighted avg.	0.97	0.97	0.97	160

The second control loop, based on bacterial interaction, processes the images in parallel and uses two images captured simultaneously by the two cameras (stereoscopic system), which are 0.5 m from the ground. These images are scaled in this case to 800×600 pixels. The projection of the bacteria i on the images is determined by equations 3 and 4.

$$\text{Left image}\begin{cases} x_p = 400 + \frac{(x_i+0.14)800}{2z_i\tan(35^\circ)} \\ y_p = 300 - \frac{(y_i)600}{2z_i\tan(30^\circ)} \end{cases} \quad (3)$$

$$\text{Right image}\begin{cases} x_p = 400 + \frac{(x_i-0.14)800}{2z_i\tan(35^\circ)} \\ y_p = 300 - \frac{(y_i)600}{2z_i\tan(30^\circ)} \end{cases} \quad (4)$$

The bacterial population (with a size of 50 agents) is initialized with random and known position distributed in the three-dimensional space able to be detected by the cameras of the robot, this is denoted by the dotted line in Figure 1. The coordinates of the bacterium i are given by the vector (x_i, y_i, z_i). Replacing these values in equations 3 and 4 gives the projections

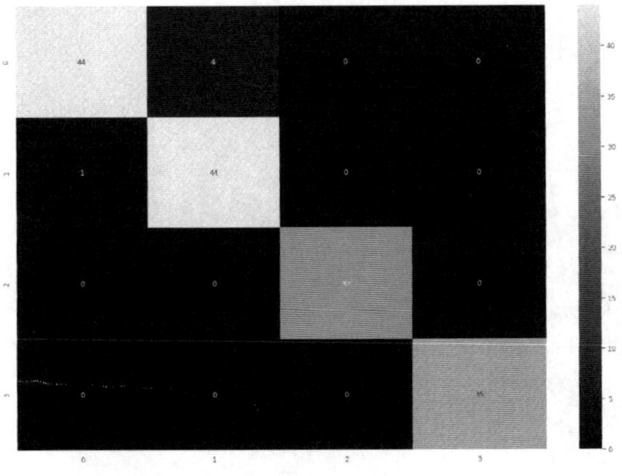

Figure 6. Confusion matrix.

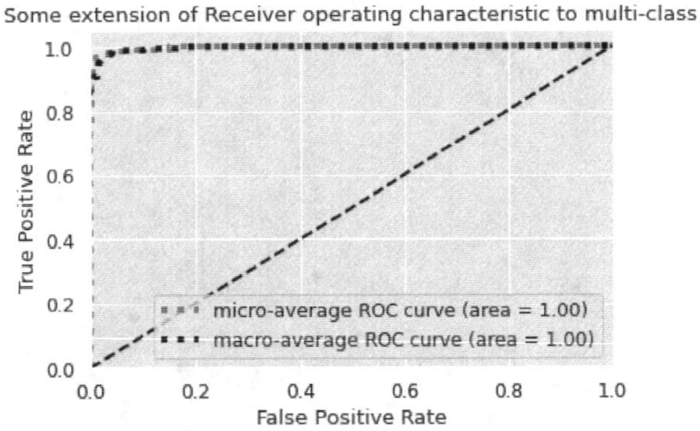

Figure 7. ROC curve (average behavior).

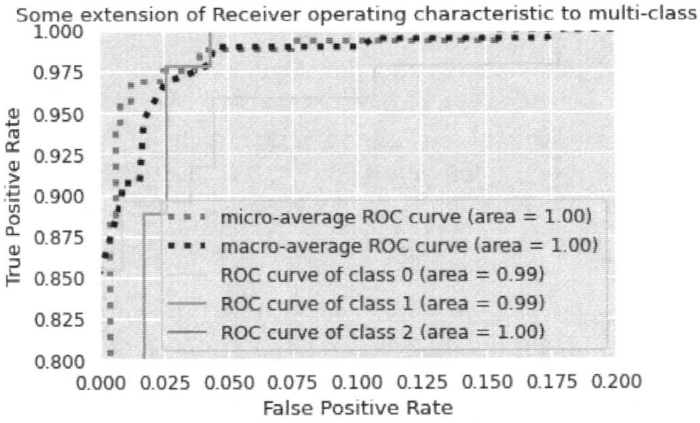

Figure 8. ROC curve (behavior by class).

in the two-dimensional images (x_p, y_p). The performance of the region neighboring bacterium i is evaluated with equation 2. If the bacterium is on the surface of an obstacle, then the neighboring pixels will be similar in both images, and the function assigns a higher value to the position of the bacterium. According to the performance value, the bacteria i navigate the state space, and after some cycles, it groups with the other bacteria on the surface of the obstacle.

The navigation strategy was evaluated on our ARMOS TurtleBot robotic platform. The VM loop has 97% accuracy in correctly identifying the environment according to the four labeled cases. On the other hand, the epipolar analysis loop with a bacterial population of 50 agents can reconstruct in great part the most outstanding elements of the obstacles according to the conditions categorized by the first loop (Figure 9). It was observed in the experiments that factors such as lighting conditions affect the performance of the strategy, however, it is an effect that also occurs in humans.

The code used to control the robot's movements, as well as the acquisition and processing of the images was written in Python [29]. The entire system ran on the robot's control unit, which was implemented on a DragonBoard 410c Development Board from Arrow Electronics, which has a Qualcomm APQ8016e 64-bit processor.

5. Conclusions

Given the need for an autonomous reactive system for our service robot, we propose a parallel control scheme based on stereoscopic vision able to operate in real-time on our platform with

Figure 9. Identification of the obstacle by bacteria.

low resource consumption. The task to be developed by the robot involves mainly autonomous navigation of long distances supported by the identification of landmarks and obstacles in the environment. Also, a low computational cost and robustness as a guarantee in the development of the task are sought. To solve the problem we implemented a control scheme based on information conformed by two parallel control loops. Our strategy is based on a loop that propagates a convolutional network and another loop that implements our bio-inspired algorithm for three-dimensional obstacle reconstruction. The two loops operate separately, but take as their information the same images captured by two cameras on the front of the robot. The neural network identifies environmental and landmark conditions, and the stereoscopic system estimates the distance to obstacles in front of the robot. This information is processed with a set of motion policies that determine the final movement of the robot. The resulting motion control scheme have several advantages over other methods that directly control the entire nonlinear system. The two control loops demonstrate accuracy ranges above 90%, allowing the robot to closely follow landmarks and avoid obstacles in 98% of cases. The results shown correspond to laboratory tests developed on our ARMOS TurtleBot robot. These tests were structured in an environment equivalent to that expected in domestic tasks, and the results fully validated the proposed strategy.

Acknowledgments
The authors wish to acknowledge and thank the Technological Faculty of the Universidad Distrital Francisco José de Caldas for the support and funding during the course of this research work.

References
[1] Decker M, Fischer M and Ott I 2017 *Robotics and Autonomous Systems* **87** 348–354 ISSN 0921-8890
[2] Winter S, Tomko M, Vasardani M, Richter K, Khoshelham K and Kalantari M 2019 *ACM Computing Surveys* **52** 1–24 ISSN 0360-0300
[3] Castañeda J and Salguero Y 2017 *Tekhnê* **14** 73–86 ISSN 1692-8407
[4] Macik M 2018 *Acta Polytechnica Hungarica* **15** 149–167 ISSN 1785-8860
[5] Courbon J, Mezouar Y and Martinet P 2009 *IEEE Transactions on Intelligent Transportation Systems* **10** 392–402
[6] Tsunekawa K, Leiva F and Ruiz J 2018 *IEEE Robotics and Automation Letters* **3** 3247–3254
[7] Bista S, Robuffo P and Chaumette F 2016 *IEEE Robotics and Automation Letters* **1** 423–430
[8] Lee K, Gibson J and Theodorou E 2020 *IEEE Robotics and Automation Letters* **5** 1207–1214

[9] Montiel H, Martínez F and Martínez F 2019 *TELKOMNIKA (Telecommunication Computing Electronics and Control)* **17** 1468 ISSN 1693-6930
[10] Ferreira M, Costa P, Rocha L and Moreira A 2016 *The International Journal of Advanced Manufacturing Technology* **85** 57–69 ISSN 0268-3768
[11] Rodrigues R, Basiri M, Aguiar A and Miraldo P 2018 *IEEE Robotics and Automation Letters* **3** 2079–2086
[12] Hongshan Y, Jiang Z, Yaonan W, Wenyan J, Mingui S and Yandong T 2014 *Sensors* **14** 10753–10782 ISSN 1424-8220
[13] Freundlich C, Zhang Y, Zhu A, Mordohai P and Zavlanos M 2017 *The International Journal of Robotics Research* **36** 1268–1285 ISSN 0278-3649
[14] Dawood Y, Ruhana K and Kamioka E 2017 *6th International Conference on Computing and Informatics (ICOCI 2017)* pp 235–242
[15] Mohamed A, Chenguang Y and Cangelosi A 2016 *4th IFAC International Conference onIntelligent Control and Automation Sciences* pp 161–168
[16] Mahammed M, Melhum A and Kochery F 2013 *International Journal of Science and Applied Information Technology* **2** 5–8 ISSN 2278-3083
[17] Martins H, Oakley I and Ventura R 2015 *Robotica* **33** 2166–2185 ISSN 0263-5747
[18] Dreier S, Savran M, Konge L and Bjerrum F 2015 *Surgical Endoscopy* **30** 11–23 ISSN 0930-2794
[19] Panjvani K, Dinh A and Wahid K 2019 *Frontiers in Plant Science* **10** 1–17 ISSN 1664-462X
[20] Solak S and Bolat E 2015 *9th International Conference on Electrical and Electronics Engineering (ELECO 2015)* pp 685–688
[21] Boonkwang S and Saiyod S 2015 *7th International Conference on Information Technology and Electrical Engineering (ICITEE 2015)* pp 232–236
[22] Bertel O, Moreno C and Toro E 2009 *Tekhnê* **6** 3–8 ISSN 1692-8407
[23] Mehta S 2008 *Computers and Electronics in Agriculture* **63** 28–37 ISSN 0168-1699
[24] Patel D, Bachani P and Shah N 2013 *International Journal of Engineering Research & Technology* **2** 2461–2464 ISSN 2278-0181
[25] Si Y, Liu G and Feng J 2015 *Computers and Electronics in Agriculture* **112** 68–74 ISSN 0168-1699
[26] Martínez F, Jacinto E and Martínez F 2020 *International Journal of Electrical and Computer Engineering* **10** 2164–2172 ISSN 2088-8708
[27] Mehrabi M, Peek E, Wuensche B and Lutteroth C 2013 *Proceedings of the Fourteenth Australasian User Interface Conference (AUIC 13)* vol 139 pp 91–100
[28] Lins R, Givigi S and Gardel P 2015 *IEEE Transactions on Instrumentation and Measurement* **64** 2950–2958 ISSN 0018-9456
[29] Moreno A and Páez C 2017 *Tekhnê* **14** 61–72 ISSN 1692-8407

Design, construction and implementation of relative humidity and temperature climatic chamber for metrology laboratory

Johan S Malpica Gutierrez , Ismael A Fernández Peña, Fernando Martínez Santa

Facultad Tecnológica, Universidad Distrital Francisco José de Caldas, Bogotá, Colombia

E-mail: jsmalpicag@correo.udistrital.edu.co,iafernadezp@correo.udistrital.edu.co, fmartinezs@udistrital.edu.co

Abstract. For the calibration of thermohygrometers it is necessary to have a means of generating temperature and relative humidity for this reason in this development we will build a climatic chamber which will generate and control these two magnitudes. For the generation of temperature a thermoresistance was used and the decrease of relative humidity was done through a silica gel trap through which the air inside the chamber is recirculated with the use of a vacuum pump. On the other hand, an ultrasonic humidity generator also known as fogger was used to increase the relative humidity. For the construction of the chamber, acrylic was used as the main material, since it is translucent and facilitates the visualization of the thermohygrometers' indications. For the control of the process variables such as temperature and relative humidity, an Arduino card was used, which through PWM pulse width modulation, the control was performed achieving a variation of 0.3 °C and 2 %RH, which will be evaluated by an accredited laboratory in the characterization of isothermal media certified by ONAC, since it is in this way that the inhomogeneity, stability and thermal load of the medium, components that affect the uncertainty of the instruments under calibration, are evaluated. It should be remembered that this development was carried out in order to create an air medium with which temperature and relative humidity measuring instruments can be calibrated in the metrology laboratory VALIDACIONES Y METROLOGIA LM S.A.S., since it is in the process of growth and one of its main objectives is to be accredited in the magnitudes of temperature and relative humidity before the accreditation body in Colombia (ONAC), which is the entity that will finally give the technical approval to release the chamber and put it into operation.

1. Introduction

The laboratory VALIDACIONES Y METROLOGÍA L&M S.A.S. is a national laboratory (Colombia) with more than six years of experience in calibration and validation of measuring instruments. Based on the growth and development of the laboratory it is necessary the creation and implementation of an air-conditioned chamber for the calibration process of thermohygrometers and meters of environmental conditions.

This article describes the design and construction of a prototype climatic chamber that was carried out in order to generate and control environmental conditions in temperature and relative humidity over a certain operating range to perform calibration tests to thermometers and hygrometers, in compliance with the current national and international technical requirements contained in the NTC-ISO/IEC 17025 standard, which establishes the standards for the management of equipment and tests in metrology laboratories. [1][2][3].

This development is based on the following objectives to achieve the Accreditation before the national accreditation body of Colombia ONAC in the calibration of measuring instruments in the magnitudes of ambient temperature and relative humidity, obtaining a measurement and calibration capacity (CMC) competitive in the market[4].

Based on the design and construction of the climatic chamber, an evaluation of its performance will be carried out based on a characterization that allows determining the behavior of the variables to be controlled according to the measurement intervals for the equipment to be calibrated, establishing the operating ranges for temperature (25 °C to 50 °C) and relative humidity (10 %HR a 90%HR).

2. Methodology

The following flow chart describes the development process of the climatic chamber.

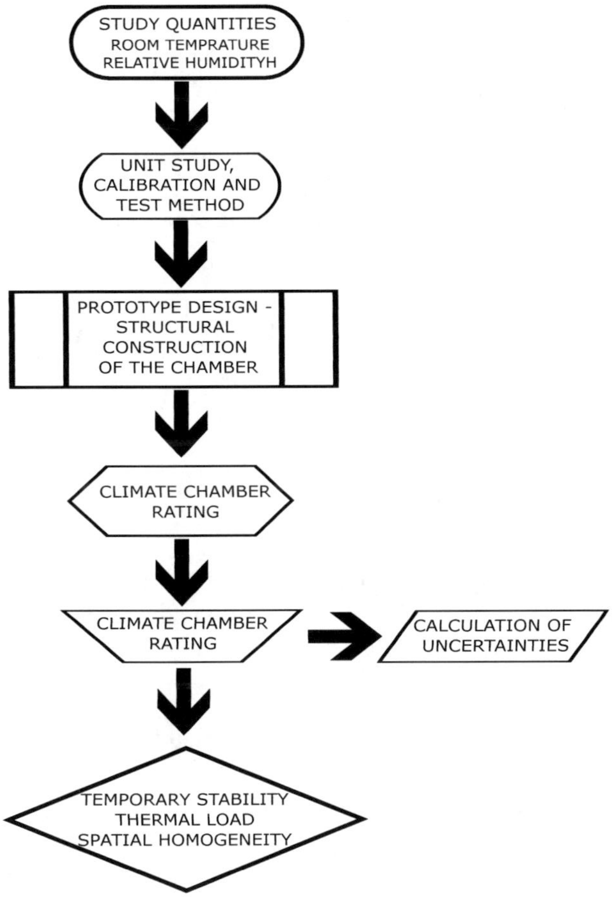

Figure 1. Development process flow chart climatic chamber.

Journal of Physics: Conference Series **2135** (2021) 012003 doi:10.1088/1742-6596/2135/1/012003

The development of this work is presented as follows. In the first stage, the development and construction of the climatic chamber is carried out, defining the basic concepts of functioning and operation. In the second stage, the control system is described by means of a block diagram to model. This control of the equipment maintain the Set Points (SP) initially established in the chamber for the ambient temperature and relative humidity. In the last stage, the chamber operation tests will be carried out in order to determine the contribution to the uncertainty that this will provide to the instruments under calibration.

3. Development and construction of the climatic chamber

he main material used for the construction of the stability chamber was acrylic since it is a material that allows to maintain the conditions due to its thermal insulating property, besides it is also possible to visualize the indication and behavior of the instruments under calibration. The design is based on a rectangular structure of dimensions 40 cm x 26 cm x 26 cm, which has a door on the right side for the entry and exit of the instruments under calibration, this same one has a gasket that isolates the internal environmental conditions from the environmental conditions of the laboratory, additionally two (2) acrylic boxes were built, which are external to the stability chamber.

The first box houses the power and control circuit, which will be explained in the Control system section. The second box houses the humidifier or fogger whose function is to change the water in the reservoir from a liquid to a gaseous state, generating humidity in the chamber according to the signal received from the controller[5]. To carry out the dehumidification process in the chamber, a vacuum pump was installed which removes the humid air from the chamber by passing it through a trap with a $CaSO4$ desiccant filter and the dry air is recirculated back into the chamber; it should be noted that the fittings and hoses through which the air circulates are totally airtight [6] [7][8].

The temperature regulation system is built by means of a resistor which receives the control signal and is activated generating an increase in temperature [9]. To reduce the temperature, the heating element is deactivated and, by recirculation of air into the chamber, the temperature is lowered.

4. Control system

The control system was made by programming in C language using free Arduino software and UNO motherboard as embedded system. The variables to be controlled will be the ambient temperature and relative humidity of the air. The program designed allows the control to react to changes in environmental conditions in the chamber and quickly compensates the values of the variables according to values preseted by activating the different actuators. The control to be applied on these variables must take into account the environmental conditions and the different thermal loads that can be generated, added or extracted from the interior of the space to be controlled.[10].

The actuators used in the process plant to control environmental conditions are:

- For heating: 110 VAC thermoresistance at 3 000 Watts of forced convection.
- For humidification: 24 VDC ultrasonic fogger with natural convection.
- For Dehumidification: 110 VAC air recirculation vacuum pump for desiccant $CaSO4$.

Figure 2. Stability chamber final result.

Stability and uniformity conditions inside the chamber are obtained by forced convection using four (4) fans strategically arranged at the corners and in opposite directions to each other.

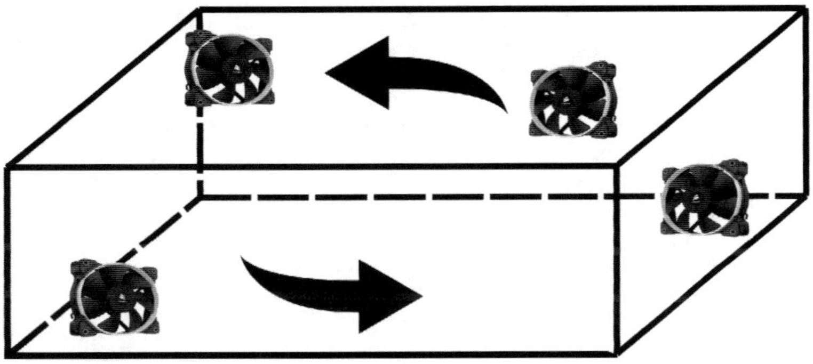

Figure 3. Forced convection air flow representation.

The control system established for the resistance thermometer is by PWM pulse width modulation ensuring low overshoot and steady state error of ± 0.1 ° C approximately, for the Fogger and pump controls are ON / OFF due to the same nature of the actuators, however as a measure of protection for the actuators of the chamber, the control system enters Standby mode when conditions are in a range set around the Set established, ± 0.3 ° C in temperature and ± 0.7 % hr for relative humidity. The control to apply on these variables must take into account the environmental conditions and the different thermal loads that can be generated, added or extracted from inside the space to be controlled.

The block diagram of the closed-loop control circuit is illustrated below.

Journal of Physics: Conference Series **2135** (2021) 012003 doi:10.1088/1742-6596/2135/1/012003

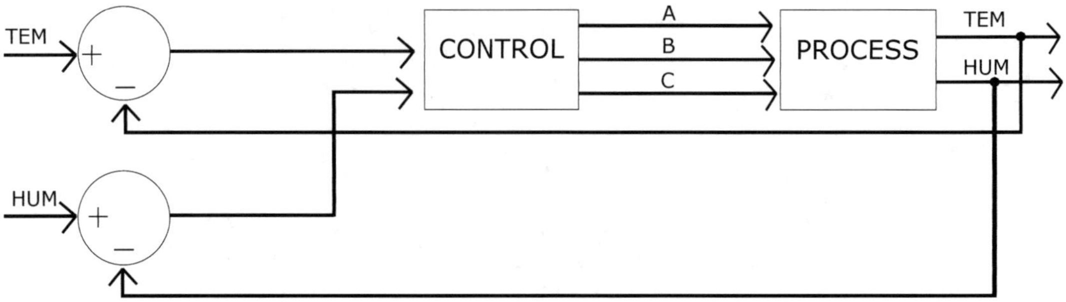

Figure 4. Representative block diagram of the control system where A. Thermo-resistance B. Fogger C. Vacuum pump

5. Functional testings

After assembling the climatic chamber, a series of functional tests were performed in order to verify if the environment is suitable for the calibration of thermometers and hygrometers, not without first remembering that it is a forced convection air environment under conditions that must be taken into account when qualifying or characterizing the equipment

Figure 5 shows the relative humidity indication vs. time when the chamber is programmed at a set point of 70 %RH, where the stabilization phase of the relative humidity inside the chamber and its behavior in the stable phase can be observed.

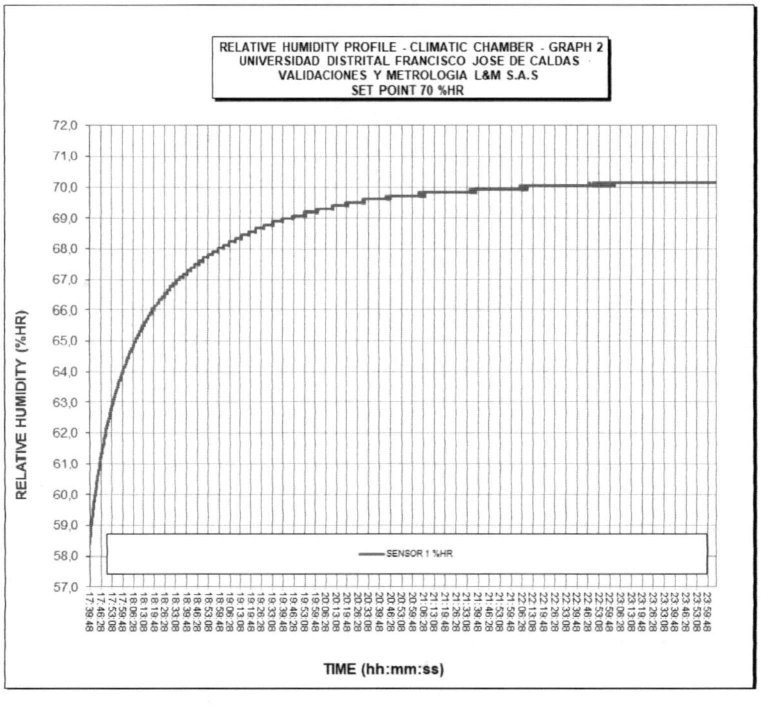

Figure 5. Relative humidity graph

Figure 6 shows the temperature vs. time indication when the chamber is programmed at a Set point of 40 °C, where the temperature stabilization phase inside the chamber and its behavior in the stable phase can be observed.

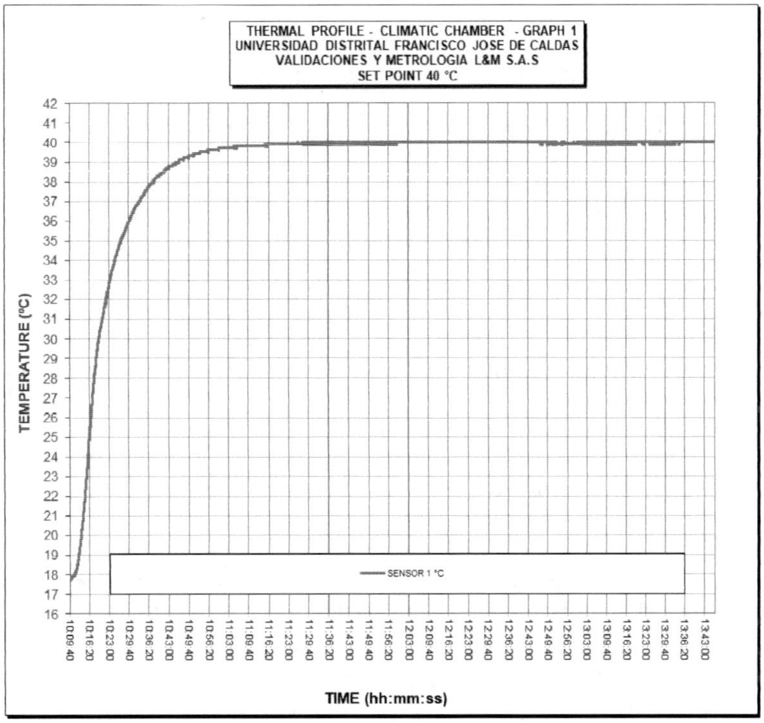

Figure 6. Temperature graph

6. Characterization of the environment

Characterization of the medium is performed in order to check the temperature deltas throughout its useful volume [11]. The temperature delta are determined from the determination of the temporal instability of the chamber, the spatial inhomogeneity, the thermal load and, if considered, the possible effect of radiation emitted by the chamber walls, but since the chamber is made of acrylic, this variable is considered negligible. Evaluating all these variables, the contribution to the uncertainty provided by the climatic chamber is determined [12].

The measured values of ambient temperature and relative humidity are the average of 10 data at each position, the ambient temperatures shown are determined using a thermometer with low emissivity.

The characterization of the equipment is carried out taking into account the useful volume of the equipment, since the distribution of the location of the measurement sensors in the X, Y and Z axes depends on it.

7. Evaluation of uncertainty

The uncertainty of measurement is known as a non-negative parameter that characterizes the dispersion of the values attributed to a measurand, from the information that is used, in other words, it is the degree of "doubt" that is had on the result of any measurement, the components of uncertainty that the camera contributes to the result of the calibration of thermohygrometers are those of stability and uniformity and this design seeks that these values are the lowest possible within reason and with the support of an accredited laboratory in characterization of isothermal media [13].

The contribution for medium uniformity corresponds to the uniformity value of the environmental conditions chamber divided into $\sqrt{3}$.

$$Uunf = \frac{Uniformity}{\sqrt{3}} \tag{1}$$

The contribution for stability of the medium corresponds to the uniformity value of the environmental conditions chamber divided into $\sqrt{12}$.

$$Uunf = \frac{Uniformity}{\sqrt{12}} \tag{2}$$

DIGITAL THERMOHIGROMETER										
MEASUREMENT INTERVAL		50 %hr								%hr
SYMBOL	UNCERTAINTY COMPONENT	DEAR (xi)	UNITS	PROBABILITY		k/DIVIDER	COEFFICIENT OF SENCIBILITY	CONTRIBUTION TO UNCERTAINTY	DEGREES OF FREEDOM	CONTRIBUTION
				KIND	DISTRIBUTION			U	Vi	%
U smp	*Pattern measurement system*	0,8	%hr	B	t-Student	2	1	0,4	200	21,73%
U RIC	*Resolution of the instrument to be calibrated*	1	%hr	B	Rectangular	3,464	1	0,2887	200	15,68%
U p	*Inhomogeneity of the environment*	1,543	%hr	B	Rectangular	2	1	0,7715	200	41,92%
Ues	*Medium stability*	0,188	%hr	B	Rectangular	2	1	0,0938	200	5,10%
U d	*Pattern drift*	0	%hr	B	Rectangular	1,732	1	0	200	0,00%
Uin	*Interpolation*	0,366	%hr	B	Rectangular	2	1	0,183	200	9,94%
U Rep	*Repeatability*	0,141	%hr	A	Normal	2	1	0,0707	9	3,84%
Umet	*Calibration method*	0,004	%hr	B	Rectangular	1	1	0,004	200	0,22%
U T	*IR resolution*	0,1	%hr	B	Normal	3,464	1	0,0289	200	1,57%
COMBINED UNCERTAINTY								0,9416		100%
EFFECTIVE DEGREES OF FREEDOM								405		
k								2	Dominant uncertainty assessment	
INCERTIDUMBRE EXPANDIDA °C								1,88		
INCERTIDUMBRE EXPANDIDA °C								0,002	2,13	
		NUMBER OF DECIMALS TO ENSURE 2 SIGNIFICANT FIGURES						1		

Figure 7. Uncertainty components associated with calibration of thermohygrometers

During the calibration process of thermohygrometers other uncertainty components are taken into account as detailed below, finally the combination of all these components multiplied by the respective coverage factor will estimate the expanded uncertainty for a confidence level of

Journal of Physics: Conference Series **2135** (2021) 012003 doi:10.1088/1742-6596/2135/1/012003

approximately %.95 [14][15].

A calibration may be expressed by a statement, a calibration function, a calibration diagram, a calibration curve or a calibration table. In some cases, it may consist of an additive or multiplicative correction of the indication with its corresponding uncertainty [16][17][18].

8. Results

To evaluate the results of the development of the climatic chamber, a calibration was requested by a laboratory accredited by the national accreditation body ONAC, and based on the results obtained in the characterization, the effectiveness of the development is evaluated.

From the calibration performed by the metrology laboratory SET Y GAD S.A.S. with accreditation code before the national accreditation body of Colombia ONAC 18 -LAC-004, the calibration certificate number HM0310-21 was issued, in which the following results were obtained.

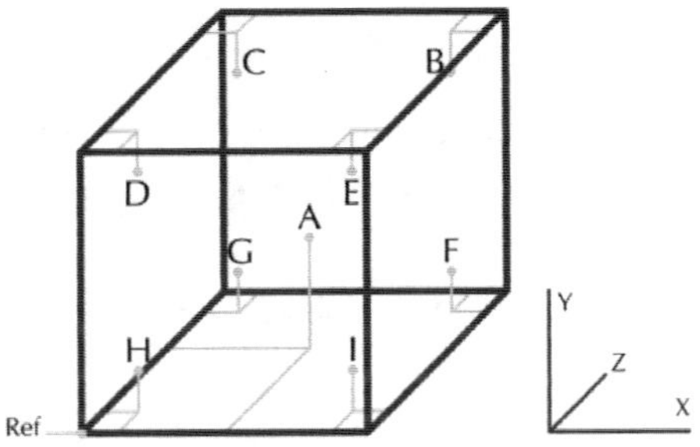

A:	12,5 cm × 7,5cm ×20,5 cm	D:	2 cm × 13cm ×9 cm	G:	2 cm × 2cm ×31 cm
B:	23 cm × 13cm ×31 cm	E:	23 cm × 13cm ×9 cm	H:	2 cm × 2cm ×9 cm
C:	2 cm × 13cm ×31 cm	F:	23 cm × 2cm ×31 cm	I:	23 cm × 2cm ×9 cm

Figure 8. Location of sensors during characterization

Table 1. Design conditions VS real conditions in the magnitude HR.

Evaluated Parameter	Design	Result of characterization	Acceptance requirements
Stability	0,5 %HR	0,194 %HR	Meets
Uniformity	2,5 %HR	1,98 %HR	Meets
Thermal load	1 %HR	0,52 %HR	Meets

In conclusion, when analyzing the stability, uniformity and thermal load data, it can be seen that the design conditions were exceeded, since in terms of temperature uniformity a value of 0.5 °C was expected and with the characterization process a value of 04030 °C was obtained, in the same way, values lower than those expected in the design stage of the climatic chamber for the calibration of thermohygrometers were obtained.

Table 2. Design conditions VS real conditions in the magnitude temperature.

Evaluated Parameter	Design	Result of characterization	Acceptance requirements
Stability	0,3 °C	0,168 °C	Meets
Uniformity	0,5 °C	0,4030 °C	Meets
Thermal load	0,1 °C	0,008 °C	Meets

9. Acknowledgments

We express our sincere gratitude to the metrology laboratory VALIDACIONES Y METROLOGÍA L&M S.A.S. for facilitating the use of the facilities and monetary resources for the development of this project; also to the ARMOS research group for their support in the construction of this work; and to Engineers Homar Alirio Lopez and Victor Alfonso Ballesteros for their support in the creation and implementation of the climatic chamber.

References

[1] Muñoz Guatibonza J A 2018 *Diseño de una cámara climatizada, automatizada, con manejo de información por telemetría y condiciones operativas controladas para pruebas de equipos y materiales* B.S. thesis Fundación Universidad de América
[2] Arango O E A 2018 *Diseño e Implementación de un prototipo de control de humedad relativa para una cámara existente en el laboratorio de metrología de variables eléctricas de la* Ph.D. thesis Universidad Tecnológica de Pereira
[3] Kalibrierdienst D 2004 *Guideline DKD-R 5-7 Calibration of climatic chambers Deutscher Kalibrierdienst (DKD)* Deutscher Kalibrierdienst DKD 7th ed
[4] Tamani B V 2007 *Obtención de modelos de procesos mediante métodos de identificación recursiva UNMSM-Facultad de Ingeniería Electrónica y Eléctrica de la Universidad Nacional Mayor de San Marcos, Lima, Perú* 27–35
[5] Fan K C, Wang H and Liu Y C 2008 *Development of a constant temperature environment chamber with high stability* vol 594 (Trans Tech Publ) pp 78–83
[6] Mahmoud E E D 2009 *Realization of relative humidity scale from 10% to 98% at 25° C Mapan* **24** 241–245
[7] Lee S W, Choi B I, Woo S B, Kim J C and Kim Y G 2019 *Development of a low-temperature low-pressure humidity chamber for calibration of radiosonde humidity sensors BIPM & IOP Publishing Ltd* **56**
[8] Cram L 1956 *Journal of scientific instruments* **33** 273
[9] Van Geel J, Bosma R, Van Wensveen J and Peruzzi A 2015 *Thermistors used in climatic chamber at high temperature and humidity International Journal of Thermophysics* **36** 569–576
[10] Cardona Gil J A, Pazos Urrea J P *et al.* 2014 *Cámara de ambiente controlado para la supervivencia de plantas e insectos Universidad Pontificia Bolivariana*
[11] Davis J B 2009 *Handbook of Stability Testing in Pharmaceutical Development* (Springer) pp 285–302
[12] WONG G and HERMAN LAM P 2005 *Qualification of Environmental Chambers Analytical Method Validation and Instrument Performance Verification* 243
[13] Aranda N G, Rocio N M, García Skabar J *et al.* 2021 Optimización de los métodos y procedimientos de calibración de termohigrómetros en el Centro Regional de Instrumentos de Buenos Aires Servicio Meteorológico Nacional Dirección Nacional de Infraestructura
[14] FENOTTI F and VASTY D 2019 *19th International Congress of Metrology (CIM2019)* (EDP Sciences) p 18003
[15] Dona M L 2010 *University" Politehnica" of Bucharest Scientific Bulletin, Series C: Electrical Engineering* **72** 197–210
[16] Vendt P D R and Leito I 2014 *2014 A method for thermal ambient tests of space technology equipment in a thermal chamber-development and validation* Master's thesis
[17] Esquivel A E 2015 *Metrología y sus aplicaciones* 2nd ed (Larousse, grupo editorial patria)
[18] Kumar S P, Shanmugasundaram N and Ganesh E 2018 *International Journal of Engineering & Technology* **7** 307–310

Optimization of an automatic voltage regulator AVR on a synchronous machine using fuzzy control

Diego N Cuesta Cuesta, Fernando Martínez Santa

Facultad Tecnológica, Universidad Distrital Francisco José de Caldas, Bogotá, Colombia

E-mail: dncuestac@correo.udistrital.edu.co, fmartinezs@udistrital.edu.co

Abstract.
The constantly developing society demands more and more electronic devices and microchips that perform vital tasks such as medical services, emergency lighting, communication systems, among others, however these are sensitive to variations and failures of the power supply, such as voltage fluctuations, voltage spikes and interference, for this reason and because of its great importance for its proper functioning must have a continuous power supply, Thus, the work shown in this article proposes to optimize the operation of automatic voltage correction devices AVR used in synchronous power generating machines whose main function is to ensure that the voltage has been constant, for this a solution is proposed based on the use of non-traditional control techniques such as fuzzy logic, For this purpose initially recognizes the relevant elements that make up the AVR system which are amplifier, exciter, generator and sensor then illustrates the mathematical block model that represents the operation of the system which is reduced to transfer function or otherwise as the relationship of the input and output signal of the system, , Then a possible classical PID proportional integral derivative proportional control is suggested with the help of the PID tools of the software Matlab ®, where the fuzzy logic inference set is programmed in three stages: first stage: input of the set of rules for voltage error correction, second stage: input of the set of rules for the voltage field output in the synchronous machine, and in a third stage: programming of the fuzzy inference sentences. Finally, the response of each control is compared and the methodology in the design of the alternative control in the synchronous machine is exposed, with the use of the software Matlab®, all this as a study of new trends in control for educational purposes, in the context of Technology in Electricity and Electronic Technology.

1. introduction

Electricity is the most widely used type of energy in the world, so it is essential to guarantee the continuity of this resource at every stage of its generation, distribution and transmission. However, the electrical components installed to supply and transfer energy make up a dynamic system subject to disturbances and other phenomena that affect the operation of the entire system [1] According to the above, the following study proposes to improve the voltage control in synchronous machines, which are affected by the adverse effects of the network, that means the disturbances of the system affect the stability and synchronism of the generators, which produces power losses [2].

Automatic Voltage Regulator (AVR) devices are currently used to correct these effects, these devices are based on classical PID controllers. [3]. However, new control trends propose the use of artificial inference algorithms such as fuzzy logic, which could become faster by stabilizing the voltage control variable and therefore be more efficient. This is developed in the *Simulink* tool and the Fuzzy Logic toolbox of Matlab®. The following sections present the methodology for the development of the AVR system based on fuzzy logic. In the first section the study of the mathematical model is presented, in the second section the block model that represents an AVR system, in the third section the classical PID auto-tuned control is developed with the toolbox pidtools of Matlab®. Then, in the fourth section, the theory for the implementation of the fuzzy control is developed and finally the two models are compared in terms of the stabilization.

The relevant components for the study of the operation of synchronous machines, as proposed by C. Millan in [4] and othe authors [5].

- Exciter: Provides direct current to the field winding of the synchronous machine, and constitutes the power stage of the control system.
- AC regulator: Processes the inputs from the sensors and provides a suitable control signal for the exciter. Signal processing employs classical regulation and stabilization techniques.
- DC regulator: Adjusts the field winding voltage to a certain reference value and allows manual control of the excitation. For special situations such as tests or automatic control failures.
- Excitation: Used to control excitation in special situations such as tests or automatic control failures.
- Field winding tension sensor: This sensor allows closing the loop of the manual tension control.
- Maximum excitation limiter: This protection prevents overheating due to excessive field winding current. Typically, this protection acts as a function of the field winding current.

Along with these elements described in the generator a voltage regulator AVR works correcting voltage fluctuations, these devices largely operate based on proportional integral derivative PID control, however this type of control does not have an inference engine that allows you to re-adjust and adapt quickly to adverse conditions, as the technique of fuzzy logic does, based on reasoning and ideas from fuzzy sets identifying the degree of membership that characterize an ideal signal and uses the set theory to correct the error and remain stable over time [6]. The latter is of great relevance considering that technology students rarely explore control techniques other than those described in traditional control theory. This paper proposes a methodology to implement fuzzy controllers applied to AVR as an alternative to classical PID, making use of all the tools available in Matlab® software.

2. Block model for AVR system

The mathematical model that represents the operation of the elements described are illustrated in figure 1 and are composed of four blocks: amplifier, exciter, generator and sensor according to F. Naeim [7].

Journal of Physics: Conference Series **2135** (2021) 012004 doi:10.1088/1742-6596/2135/1/012004

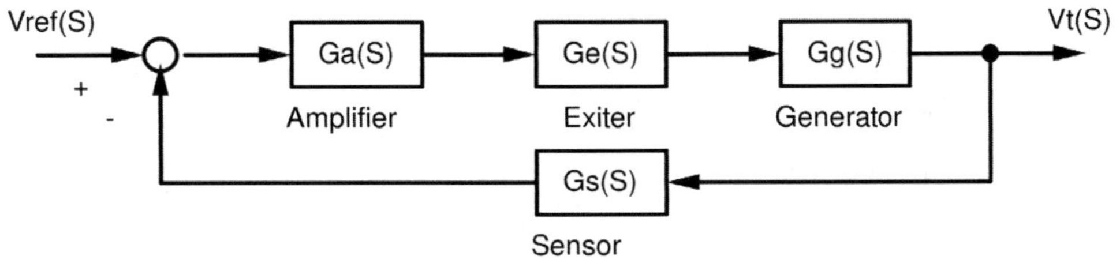

Figure 1. AVR device block model.

Amplifier Model: The transfer function of the amplifier model is:

$$\frac{V_t(s)}{V_r(s)} = \frac{KA(s)}{1 + TA(s)} \tag{1}$$

Where, de velue of KA is a constant in the range of 10 to 400 and the value of TA is in the range off 0.02 to 0.1s, a near value for KA is 5 and TA 0.1 s.

Exciter model: the transfer function of the exciter model is:

$$\frac{V_t(s)}{V_r(s)} = \frac{KE(s)}{1 + TA(s)} \tag{2}$$

Where, the constant value KE is in the range of 1 to 200 and the constant value of the amplifier time TE in the range of 0.5 to 1s. KE is set to 1 and TE is set to 0.4s for this work.

Generator Model: The transfer function of the generator model is:

$$\frac{V_t(s)}{V_r(s)} = \frac{KG(s)}{1 + TG(s)} \tag{3}$$

Here the constants depend on the load; the value of the constant KG varies between 0.7 to 1 and the generator time constant TG in the range of 1 to 2s this represents the full load on the machine. Then KG was set to 1 and TG to 1s.

Sensor model: The transfer function of the sensor model is:

$$\frac{V_t(s)}{V_r(s)} = \frac{KR(s)}{1 + TR(s)} \tag{4}$$

Where, the gain KR is generally kept at 1 and the time constant TR is very small, ranging from 0.01 to 0.06s. KR was set to 1 and TR was set to 0.01s.

3. Transfer function

By reducing the blocks proposed in the previous section, the coefficients describing the amplifier, exciter, generator and sensor are replaced by the coefficients a, b, c, d for ease of subsequent calculations, so the new transfer function is given by transfer function is given by:

$$H(s) = \frac{a_1(s)}{as^4 + bs^3 + cs^2 + ds + 1} \tag{5}$$

where

$a_1 = KA \cdot KE \cdot KG \cdot KR$

$a = TA \cdot TE \cdot TG \cdot TR$

$b = TA \cdot TE(TG + TR + TG) + TG \cdot TR(TA + TE)$

$c = TA \cdot TE + TR + TG + TR \cdot TG(TA + TE)$

$d = TR + TA + TG + TE$

According to this relationship, the values for each constant representing the behavior of the system are summarized in Table 1.

Table 1. Constants for the AVR automatic regulator model [7].

KA	KR	KE	KG	TA	TR	TE	TG
5	1	1	1	0.1	0.05	0.4	1

Replacing the values in Table 1 in equation (5) gives the transfer function representing the AVR system:

$$H(S) = \frac{5}{0.002s^4 + 0.067s^3 + 0.115s^2 + 1.55s + 1} \tag{6}$$

Next, the initial behavior of the system with respect to time is plotted in the Matlab® software, from which the following initial overshoot values of 2.52s, settling time 9.61 S and rise time 2.92 s are obtained, see figure 2 [8].

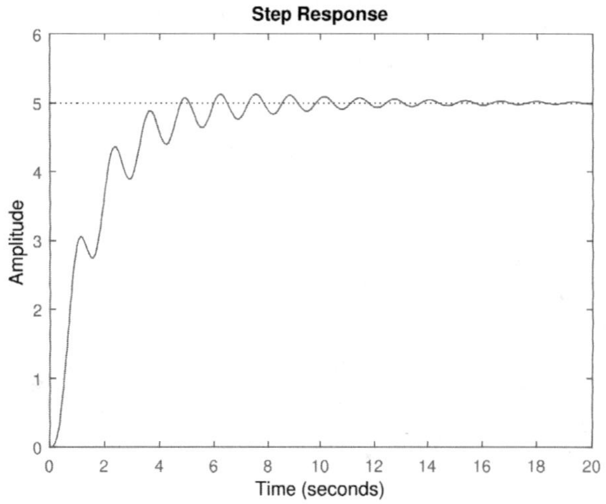

Figure 2. Plant response for the AVR system.

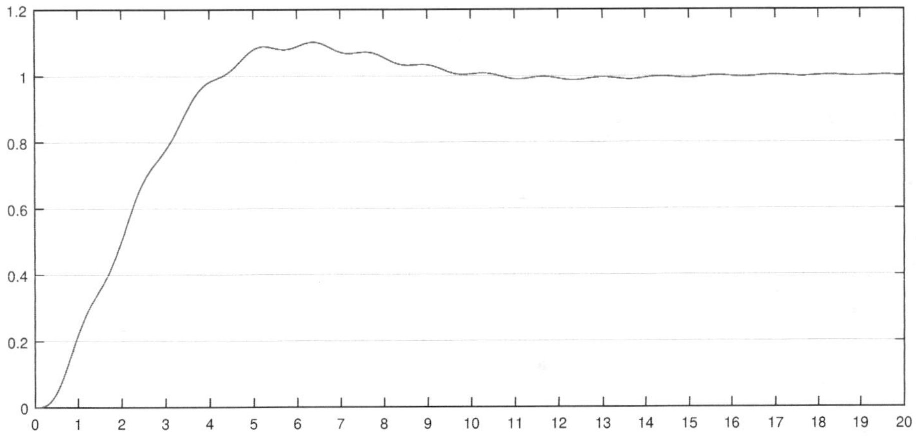

Figure 3. Plant response with PID control.

4. Classic PID control

Once the initial behavior of the AVR device is recognized, the PID control is auto-tuned by means of the pidtools toolbox of Matlab® thus achieving the stabilization of the system response, as shown in figure 3 [9].

Therefore the control constants after autotuning would be proportional constant $kp = 0,029$ integral constant $ki = 0,011$ constant derivative $kd = 0$.

5. Fuzzy PID Controller

In the first instance, the set of input and error correction rules for the variable to be optimized "voltage" is established using the Matlab® fuzzy logic toolbox, where the system is modeled by means of simple logic rules [10]. Subsequently, these rules are implemented in a fuzzy inference engine (see table 2 and figure 4).

Table 2. Set of rules for voltage error control.

If the voltage error is	Then the field voltage is
Very Negative	Very Small
Negative	Small
Normal	Medium
Positive	Large
Very Positive	Very Large

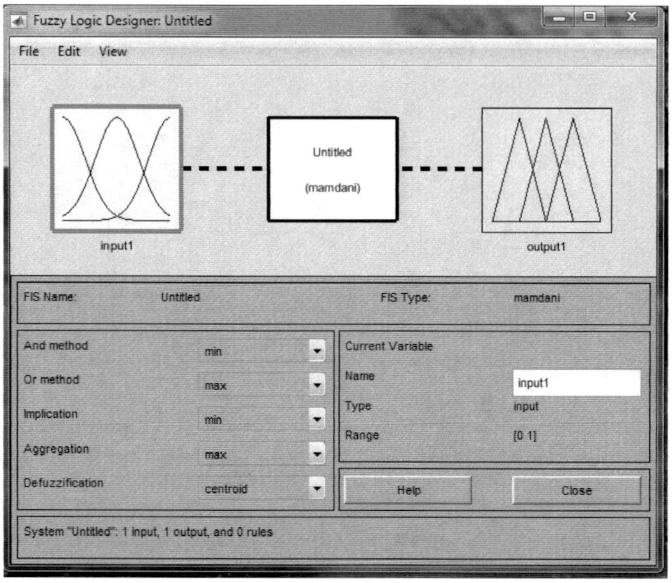

Figure 4. Error rule set programming in Matlab® Fuzzy Logic.

Once the voltage error rules are established, the fuzzy control is programmed in three stages: first stage: input of the set of rules for voltage error correction, second stage: input of the set of rules for the voltage field output in the synchronous machine, and finally the programming of the fuzzy inference statements [11].

First stage: Figure 5 illustrates the ranges for the fuzzy rules where, MN (Very Negative), NE (Negative), no (Normal), PO (Positive), MP very positive [12].

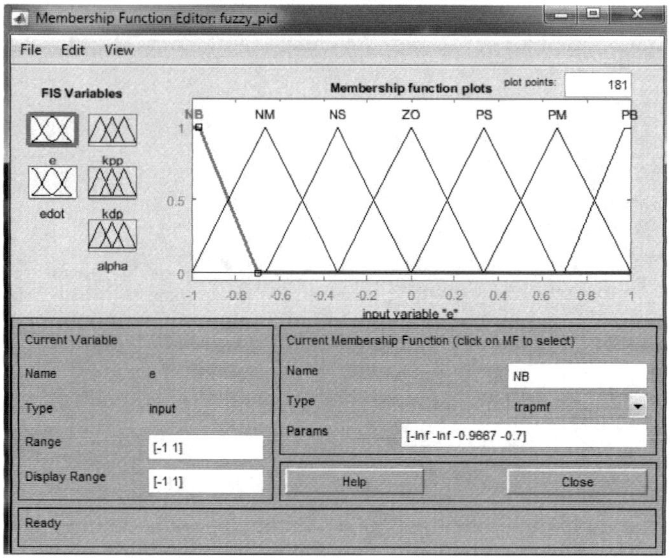

Figure 5. Tension error as a function of the proposed fuzzy rules.

Second stage: Figure 6 shows the input ranges for the field voltage of the synchronous machine where MP (very small), PE (small), ME (medium), Gr (large), MG (very large) [13].

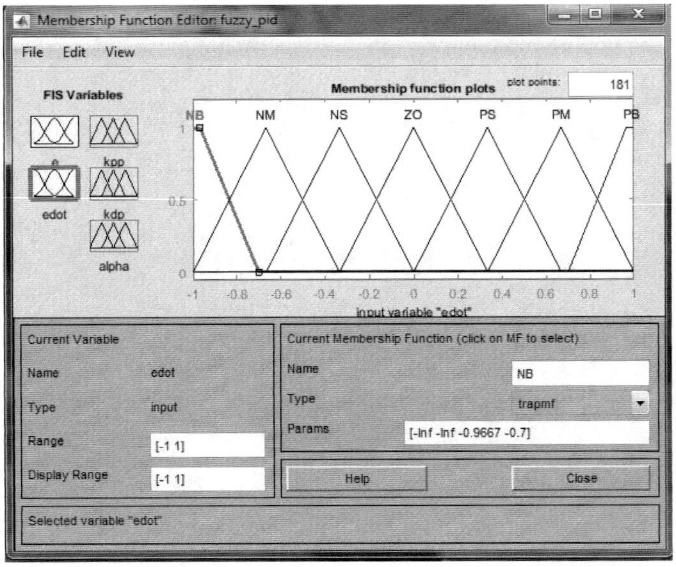

Figure 6. Field voltage output as a function of fuzzy rules.

Third stage: Figure 7 illustrates the input of the fuzzy control sentences, where the error voltage variable and field voltage are asigned [14].

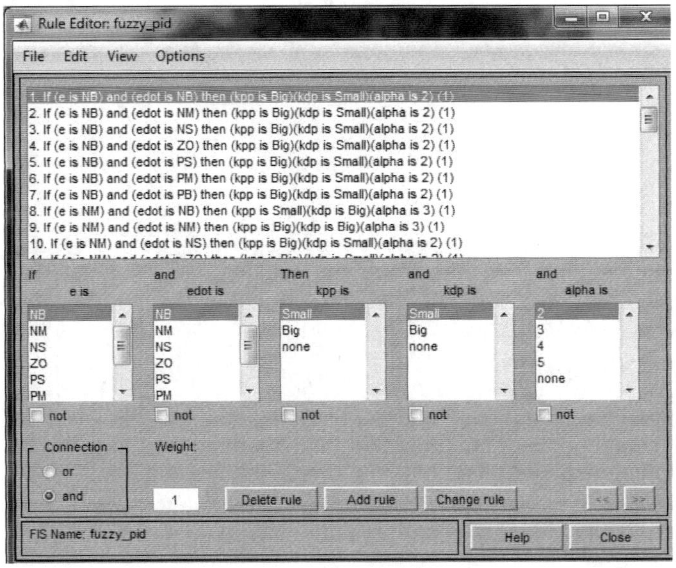

Figure 7. Programming of fuzzy inference rules.

Once the fuzzy control is programmed, the blocks are created in *Simulink* of Matlab® with the properties of the proposed fuzzy control and in parallel the PID control, in order to graphically compare the responses of each case, as shown in figure 8 [15] [16].

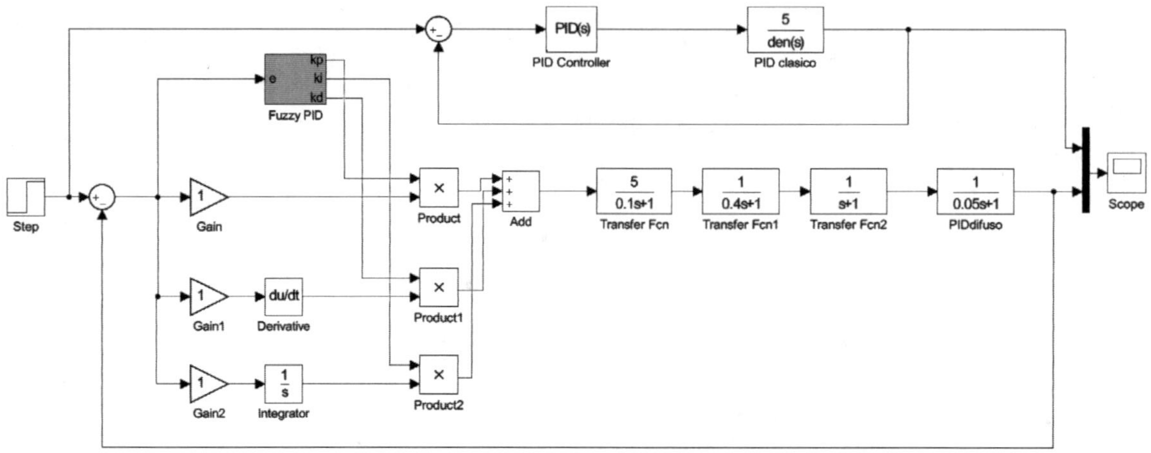

Figure 8. Parallel of Fuzzy Control and Classic Pid Control.

6. Results and discussion

It is validated that the proposed fuzzy control can represent an improvement in both the rise time and over oscillation time (see figure 9), since this response has a stabilization time of 6 seconds before the classical PID, this means that the machine will come out faster from a sub-transient effect, which can represent an improvement in the useful life of the synchronous machine. On the other hand, the steady state error is minimized, which in turn can improve the power quality indexes.

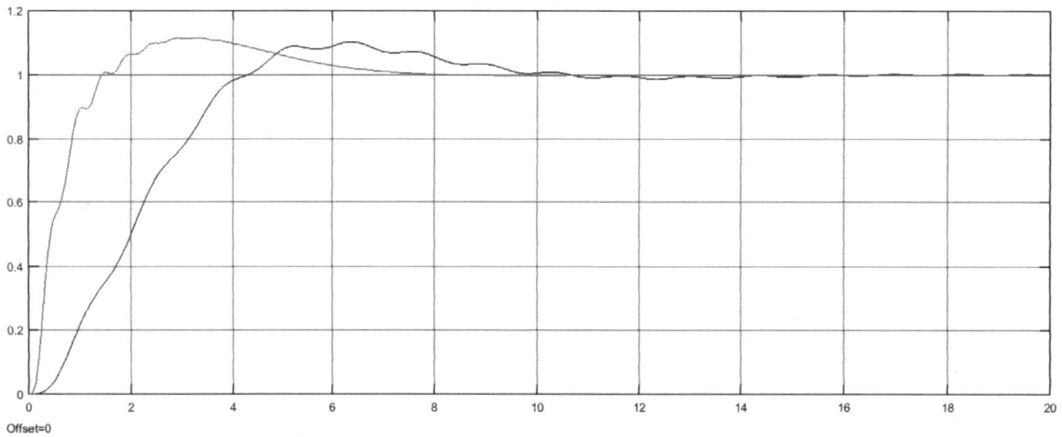

Figure 9. Responses fuzzy PID control in red compared to classical PID control in blue.

The study of the mathematical model of the synchronous machine, through the use of Matlab® *Simulink*, illustrates a practical way for academic purposes to validate the responses of the designs based on alternative controllers to PID, thus encouraging technology and engineering students to explore alternative techniques to conventional control applied to electrical machines.

As future work, it is proposed to make use of optimal control techniques on the traditional PID controller and optimization techniques on the fuzzy controller proposed (using alternative algorithms such as genetic algorithms), in order to make a comparison of the best possible results for both the PID and the fuzzy PID, using the same tools used in the development of this work.

References

[1] Posso F 2002 *Geoenseñanza* **7** 54–73
[2] Carrión Galarza D, Quinteros Flores J F and Masache Almeida P A 2021 *Revista de ID tecnológico* **17** 12
[3] Herredor C M and jaén A V 2016 Ph.D. thesis Escuela Técnica Superior de Ingeniería
[4] Herredor C M and jaén A V 2016 Ph.D. thesis Escuela Técnica Superior de Ingeniería
[5] Alzate A, Escobar A and Osorio J 2009 *Scientia et Technica* **42** 321
[6] Guzmán D and Castaño M 2006 *Ciencia y Tecnología* **24** 88
[7] Lameres B Design and implementation of a fuzzy logic based voltage controller for voltage regulation of a synchronous generator
[8] Barrera D, Suarez O and Pardo A 2019 Diseño de controladores fuzzy para el control de la excitatriz en una máquina sincrónica
[9] Reyes Sotter A M and Tobón Bayona M A 2016 Ph.D. thesis Escuela Técnica Superior de Ingeniería
[10] Blanco A 2009 Ph.D. thesis Universidad técnica de Oruro
[11] j Morales and Carbonel A 2005 *Revista Colombiana de Tecnologías de Avanzada* 6 **2** 41
[12] Eduardo C, De Vito E L *et al.* 2006 *Revista Americana de Medicina Respiratoria* **6** 126–136
[13] Puicón P A P, Alvarez A M F *et al.* 2021 *Revista UNIDA Científica* **5** 1–8
[14] Luna Ramos D *et al.* 2021 *Control de un robot móvil para mapeo con lógica difusa* B.S. thesis
[15] Entenza Boggiano V A, Argüelles Cortés L and Dorta Enriquez D D E 2021
[16] Alulema D, Andino C, Rodriguéz D, Proaño V, Ibarra A, Morocho D, Tapia F and Alulema V 2017 *2017 13th International Conference on Natural Computation, Fuzzy Systems and Knowledge Discovery (ICNC-FSKD)* (IEEE) pp 1271–1276

Ångström-Prescott Models of Solar radiation over Earth surface

Gloria Marina Bautista-Rodríguez[1], **César Alexander Chacón-Cardona**[2],

Facultad Tecnológica, Universidad Distrital Francisco José de Caldas, Cl. 68d Bis ASur No. 49F - 70, Bogotá-Colombia

E-mail: [1] gmbautistar@correo.udistrital.edu.co, [2] cachaconc@udistrital.edu.co

Abstract. The current energy demand increases every day and the deficit of electricity generation in Colombia due to technical aspects such as the lack of generation infrastructure, cost overruns in the construction projects of new plants and environmental variability as a consequence of climate change, entail a high risk of energy rationing; for this reason some of the users, whether industrial or residential, have decided to seek multiple alternatives to reduce the costs of their billing in interconnected areas and even supply their energy needs.

Among these alternatives is found solar energy in its thermal and photovoltaic use, a resource that must be characterized in order to determine the technical and budgetary aspects involved in the implementation of profitable energy generation projects. This article seeks to provide technical tools for the design and implementation of generation projects based on solar energy, providing physical-mathematical approach based in the classical Ångström-Prescott model that allow engineers and researchers to model the global solar radiation on the Earth's surface in order to maximize the use of this resource in a specific geographical location.

Although there are different ways to characterize the solar resource on the earth's surface, this paper presents a review study focused on the most important global solar radiation estimation models present in the literature.

1. Introduction

The medium-term depletion of traditional energy sources (fossil fuels) and the need to develop technologies whose impact on the environment is progressively lower, is the motive of the presented article. The increase in the cost of energy resources and the problems related to global change are forcing researchers to focus their efforts on the search for alternative sources of energy to replace fossil fuels with less impact on the environment [1].

The use of the solar resource allows the generation of electricity with a substantial reduction of environmental impacts, without greenhouse gas emissions, with fewer changes in the habitat, without noise pollution, [2], with a distributed generation close to consumers. This complementary generation is very beneficial for the community, especially for the non-interconnected zones that represent 52% of the territory of Colombia, (government report [3]) and in which service delivery may be non-existent or intermittent with low levels of reliability [4],

In other fields, applying an alternative energy model favors not only the provision of household services or public lighting, but also education [5], in the improvement of hospitals and even in public health, as solar radiation is directly related to both positive and negative effects on human beings, like sicknesses related with UV exposure or lack of vitamin D in cases of avoiding sun [6]. In agriculture, the effect of solar radiation on the quality of products, [7], or in the multiple studies carried out by agrarian associations, where linear correlations between global solar radiation and solar brightness were performed over 17 weather stations, with satisfactory results [8].

In order to take advantage of solar radiation, it is necessary to correctly characterize the energy resource, which can be done mainly through: direct measurements by means of properly calibrated radiometric instrumentation, large-scale determination by means of satellite images or the use of radiation atlases [9], and physical models. This approach provides more accurate information than that obtained by satellite and does not require the use of expensive instrumentation; it is based on the physical-mathematical connections between meteorological observations and solar irradiance at the location to be characterized.

This representation provides more accurate information than that obtained by satellite and does not require the use of expensive instrumentation; it is based on the physical-mathematical relationships between meteorological observations and solar irradiance at the site to be characterized. In addition to allowing the sizing of systems (photovoltaic and thermal) for energy use, the models for the determination of surface solar radiation are an effective tool to fill the gaps in existing historical series [10], determination of atmospheric pollutants [11], determination of protection indicators such as UV index of exposition [12] and measurements of atmospheric turbidity [11] between other applications.

2. Solar Radiation fundamental physics

The sun is the largest source of energy for planet Earth, it provides directly or indirectly all the energy that sustains life, all food and fuel ultimately comes from plants that use the sun's energy to convert it into food through photosynthesis. Energy from the sun is generated in the solar core, in a thermonuclear fusion process in which hydrogen is transformed into helium. In this process, every second an approximate mass of 4.4 million tons is transformed into energy, which radiates 3.96×10^{26}

The energy coming from the Sun is the engine of the dynamics of the atmosphere, the ocean, the climate and the terrestrial biosphere. Solar energy reaches the Earth's surface and is converted into heat due to differences in the absorption properties of the surface. This heating is not homogeneous, so that different temperatures induce movements within the atmosphere and the oceans. This energy is an essential factor in the global biogeochemical cycles, being the primary factor in the processes of water evaporation and photosynthesis in plants [13].

Solar radiation reaching the earth's surface is attenuated in its intensity by various energetic processes that occur along its path through the earth's atmosphere. These processes are [14]:

(i) Selective absorption by gases and by water vapor in the atmosphere. In this process, the gaseous molecules present in the atmospheric medium, due to their spectra, are capable of retaining electromagnetic radiation with energies of the order of differences between electronic, vibrational and rotational energy levels.

(ii) Molecular scattering, also due to gases and water vapor, occurs when the spherical particle (air molecule) has a size much smaller than the wavelength of the incident radiation. This solution was derived in the 19th century by Lord Rayleigh, which is particularly useful for the study of the scattering of solar radiation by air molecules and it is responsable of the blue color of the sky in noon and the red color in sunrise and the afternoon.

(iii) Dispersion and absorption by aerosols or atmospheric turbidity. When the particle size is at the order of the wavelength of the incident radiation, the solution of the wave equation

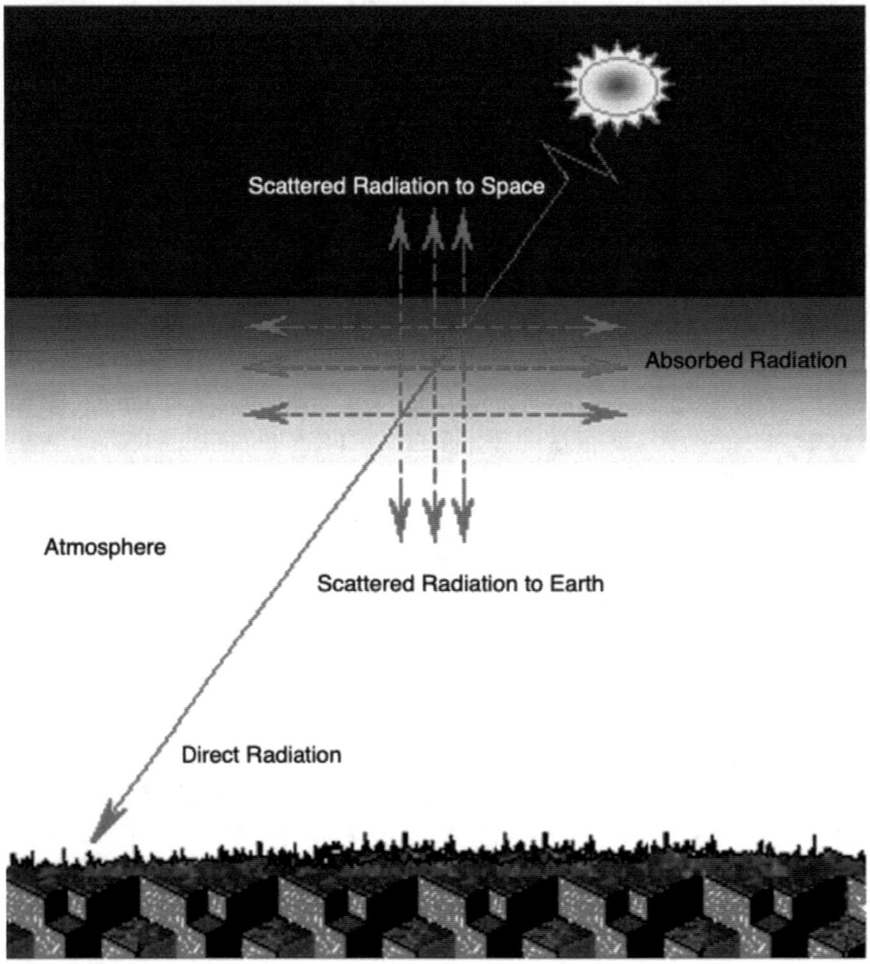

Figure 1. Solar radiation components

is extremely complex. Only at the beginning of the 20th century, Gustav Mie successfully obtained the solution, where Rayleigh's solution is only one case of Mie's theory.

When solar radiation enters the earth's atmosphere, a portion of the incident energy is attenuated by scattering and absorption. The portion of incident radiation that reaches the ground directly in line from the solar disk is called Direct Radiation, the portion that reaches the ground after being affected by attenuation processes is called Diffuse Radiation [15]. A portion of the scattered radiation returns to space (Backward Scattering) and another portion reaches the surface (Forward Scattering) See Figure 1.

The radiation reaching the earth's surface, the sum of the direct and diffuse components, is called global radiation and is measured in dimensions of power per unit area $\left[W/m^2\right]$

The first models for the determination of global radiation were based on solar brightness, among them the most important in the literature is the Ångström-Prescott model [16], this model after its creation has had different interpretation of several authors who have made adjustments or improvements that allow them to use it in different parts of the world for various tasks. [17, 18]

3. Ångström-Prescott Model

It is a linear model based on the relationship between the global radiation measured over the surface, and the solar brightness, the number of hours where direct radiation can be produce shade. It determines the the slope and intercept, which are obtained from an estimated regression of data from historical series of the two physical variables [19]. The relation between variables can be described by the equation:

$$\frac{H}{H_0} = a + b\left(\frac{S}{S_0}\right) \tag{1}$$

where: H is the global solar radiation measured at the station; H_0 is the global extraterrestrial solar radiation received in a horizontal plane for a given day over the desired location; S is the number of hours of solar brightness measured at the surface; S_0 is the astronomical day length obtained from the movement of the Earth around the sun; a and b are the regression coefficients of a simple linear model, determined by least squares. H_0 and S_0 were obtained considering the geographical position of the site, the solar declination, the hour angle for sunrise and the solar constant. Since the model is based on the translational motion of the Earth, the solar astronomy equations are necessary for the determination of H_0 and S_0. From the angular position of the Earth moving around the Sun [9]:

$$\alpha = \frac{2\pi(nd-1)}{365} \tag{2}$$

where: nd= nd is the number of the day in the year. The solar declination, the angle between the apparent position of Sun and the plane of the Earth's equator is expressed in terms of the Fourier Series:

$$\delta = 0.006918 - 0.399912\cos\alpha + 0.070257\sin\alpha - 0.006758\cos 2\alpha + 0.0000907\sin 2\alpha - 0.002697\cos 3\alpha + 0.00148\sin 3\alpha \tag{3}$$

With the latitude φ and the declination the astronomical length of the day is calculated:

$$So = \left(\frac{2}{15}\right)[\cos^{-1}(-\tan\varphi\tan\delta)] \tag{4}$$

Now is posible to calculate the radiation at the upper boundary of the atmosphere:

$$Ho = \left(\frac{24}{\pi}\right) I_0 E_0 \left(\cos\varphi\cos\delta\sin\omega + \frac{2\pi\omega}{360^\circ}\sin\varphi\sin\delta\right) \tag{5}$$

where $I_0 = 1367W/m^2$ is the solar constant, E_0 is the correction for eccentricity of the orbit:

$$E_0 = 1.00011 + 0.034221\cos\alpha + 0.00128\sin\alpha + 0.000719\cos 2\alpha + 0.000077\sin 2\alpha \tag{6}$$

And ω is the hour angle for the sunrise for every day:

$$\omega = \cos^{-1}(-\tan\varphi\tan\delta) \tag{7}$$

This procedure was used succesfully by the author César Chacón in the 2005 Solar radiation Atlas of Colombia [9] and was again used in the 2015 interactive atlas version available at IDEAM's website [20].

Different corrections have been developed on this model, each estimation involves different variables, such as the latitude present in the Glower and McCulloch model [21]; each of these with the possibility of delivering a model for various parts of the world as in the Page model [22], in which the variables a and b vary leaving (S/So) free. Some of the models that have been obtained from the corrections to the Ångström-Prescott model are listed as follows.

3.1. Glower y McCulloch Model:

In addition to the variables already established in the Ångström-Prescott model, this new model takes into account the latitude of the site, which is symbolized as φ and is valid as long as it is less than 60° [21], and is represented as follows:

$$\frac{H}{Ho} = a(\cos\varphi) + b\left(\frac{S}{So}\right) \tag{8}$$

3.2. Page Model:

This is considered the model that can be used in any part of the world, however, it should be taken into account that it was made for a latitude of 40° [17],[22], therefore it is better to recalculate the values for the coefficients of a and b, this model is represented in the following way:

$$\frac{H}{Ho} = 0.23 + 0.48\left(\frac{S}{So}\right) \tag{9}$$

where the coefficients a and b have an established value.

3.3. Bahel Model:

It is a correlation based between hours of bright sunshine and data from 48 weather stations located around the world, under different geographical locations and meteorological conditions, [23] and is represented as follows:

$$\frac{H}{Ho} = a + b\left(\frac{S}{So}\right) - c\left(\frac{S}{So}\right)^2 + d\left(\frac{S}{So}\right)^3 \tag{10}$$

where a, b, c and d are the product of regressions, with a set value of 0.16, 0.87, 0.61 and 0.34 respectively.

3.4. Benson Model

Based on Ångström-Prescott, 2 different coefficients are proposed based on the similarities of the climatic conditions of the months of the year, as follows: January-March, October-December, and April-September [24], thus:

- January-March, October-December

$$\frac{H}{Ho} = 0.18 + 0.60\left(\frac{S}{So}\right) \tag{11}$$

- April-September

$$\frac{H}{Ho} = 0.24 + 0.53\left(\frac{S}{So}\right) \tag{12}$$

3.5. Gopinathan Model:

It takes into account the latitude, the function of the elevation called Z in km and the possible percentage of insolation [25], it is based on the Ångström-Prescott model, it gives three possible functions one for a and one for b:

$$a/b = A + B * Z + C\left(\frac{S}{So}\right) \tag{13}$$

$$a = 0.265 + 0.07 * Z - 0.135\left(\frac{S}{So}\right) \tag{14}$$

$$b = 0.401 - 0.108 * Z + 0.325\left(\frac{S}{So}\right) \tag{15}$$

In addition, it provides two possible correlations found thanks to measured data of radiation and sunshine duration in 14 different locations, which can be used in other places with similar geographic and climatic conditions:

$$a = -0.0309 + 0.539 * (\cos\varphi) - 0.0693 * Z + 0.290\left(\frac{S}{So}\right) \tag{16}$$

$$b = 1.527 - 1.027 * (\cos\varphi) + 0.0926 * Z - 0.359\left(\frac{S}{So}\right) \tag{17}$$

In this form:

$$\frac{H}{Ho} = \left[-0.0309 + 0.539 * (\cos\varphi) - 0.0693 * Z + 0.290\left(\frac{S}{So}\right)\right] + \left[1.527 - 1.027 * (\cos\varphi) + 0.0926 * Z - 0.359\left(\frac{S}{So}\right)\right] * \left(\frac{S}{So}\right) \tag{18}$$

where Z the height upper sea level and φ the geographic latitude.

3.6. Zabara Model:

It is a monthly correlation that relates the Ångström-Prescott model as a third order function, thus providing a coefficient for a and one for b [26], these coefficients are obtained from the study of 7 meteorological stations in Greece which are applied to the Page model formula, which can be seen in the equation (9).

$$a = 0.395 - 1.247\left(\frac{S}{So}\right) + 2.680\left(\frac{S}{So}\right)^2 - 1.674\left(\frac{S}{So}\right)^3 \tag{19}$$

$$b = 0.395 + 1.384\left(\frac{S}{So}\right) - 3.249\left(\frac{S}{So}\right)^2 + 2.055\left(\frac{S}{So}\right)^3 \tag{20}$$

3.7. O.O. Ajayi, O.D. Ohijeagbon, C.E. Nwadialo, Olumide Olasope Model:

This is a new model created for the country of Nigeria in which 24 years of daily data from different stations were used and in which variables such as: the latitude of the place, the average daily relative humidity, the daily duration of sunlight, the maximum daily temperature and the cosine of the number of days are implemented; using this, five equations were proposed from which their coefficients were derived [27]. After doing this it was found that the model with the best performance was Model 3, which is expressed as follows:

$$H = a * (\cos\varphi) + b(\cos nd) + c * T_{max} + d\left(\frac{S}{So}\right) + f\left(\frac{T_{max}}{R.H}\right) + g\left(\frac{T_{max}}{R.H}\right)^2 + h\left(\frac{T_{max}}{R.H}\right)^3 + i * (\cos\varphi) * (\cos nd) + j\left(\frac{T_{max}}{(\cos\varphi)}\right) + k(\cos^2 nd) + l \quad (21)$$

where: H is the surface radiation, φ is the geographic latitude, S are the brightness hours at day, So astronomical day duration, T_{max} is the daily maximum temperature in degrees Celsius, nd is the number of Julian day, $R.H$ is the daily Relativity Humidity with $a, b, c, d, e, f, g, h, i, j, k, l$ the correlation coefficients.

3.8. Models based on Temperature

This kind of models have an advantage over the models based on sunlight in that the variable it uses, i.e. temperature, can be the only data available in the place where the modeling is required, because it is easy to obtain. These models assume that the difference between the maximum temperature (T_{max}) and the minimum temperature (T_{min}) is directly related to the extraterrestrial solar radiation received at the surface, since the temperature variable includes radiation, humidity, cloudiness, latitude and topography of the study site, see [17] , like the next model the Bristow-Campbell model.

3.9. Bristow Campbell Model:

This model relates a simple equation suggesting the use of the transmissivity of the atmosphere, i.e. the fraction of incident radiation passing through a sample of material, in this case the atmosphere. [28].

$$\frac{H}{Ho} = a[1 - exp(-b * \triangle T^c)] \quad (22)$$

where: a is the transmitivity of the atmosphere, b and c are constants of every place, and ΔT is the difference between T_{max} and T_{min}.

3.10. Hargreaves-Samani Model:

They propose a simple equation based on the temperature difference [29].

$$\frac{H}{Ho} = a(T_{max} - T_{min})^{0.5} \quad (23)$$

where: a is an adjustment factor, recommended by the author to be 0.16 for inland regions and 0.19 for coastal regions.T_{max} the maximum temperature and T_{min} the minimum temperature.

3.11. Hunt Model:

It takes the Hargreaves-Samani model and adds a coefficient b, the intercept, and in addition to adding Ho as a multiplication factor [30]

$$\frac{H}{Ho} = a(T_{max} - T_{min})^{0.5} * Ho + b \quad (24)$$

where: a y b are adjustment coefficients, which relate the atmospheric conditions of the site and Ho is the solar constant in units of $[KWh/m2/day]$

3.12. Annandale Model:
This model was derived from the Hargreaves-Samani model, to which a correction factor is applied to the a parameter to account for the effects of reduced altitude and atmospheric thickness. [31].

$$\frac{H}{Ho} = a(1 + 2.7 * 10^{-5} * Z) * (T_{max} - T_{min})^{0.5} \tag{25}$$

where: a is an adjustment parameter, Z is the height over sea level.

3.13. Allen Model:
I take as a basis the work done by Hargreaves-Samani but suggested an improvement that includes a factor that allows self-calibration of the model, in which the atmospheric pressure of the site is included ,[32].

$$\frac{H}{Ho} = Kr(T_{max} - T_{min})^{0.5} \tag{26}$$

Kr depends of average atmospheric pressure in the site, PS which is measured in $[kPa]$, $Po = 101, 3Kpa$ is the atmospheric pressure at sea level and kra is an empirical coefficient equal to 0.17 for inland regions or equal to 0.20 for coastal regions.

$$Kr = kra\left(\frac{PS}{Po}\right)^{0.5} \tag{27}$$

Thanks to the previously developed models or the so-called classical models, improvements have been made or comparisons have been made with different expressions, which has led to the creation of new models, more recent expressions that have been treated in computer programs that speed up and develop the analysis in a faster and more efficient way, as is the case of New Temperature-based Models for Predicting Global Solar Radiation presented below:

3.14. Model of Gasser E. Hassan, M. Elsayed Youssef, Zahraa E. Mohamed, Mohamed A. Ali, Ahmed A. Hanafy:
This model was obtained from the creation of new models based on temperature for ten locations around Egypt taking the data measured from New Borg El Arab, such data were the solar radiation and temperature data for 20 years from 1983 to 2004, taken from the NASA surface meteorology and solar energy website; to obtain this model 17 new models were proposed, which yielded results that were compared with models obtained from classical literature, such as the Allen models which can be seen in equation (26), Goodin et al. and Annandale which can be seen in equation (25), in addition to this it can be observed the capacity to estimate the solar radiation in any place of the world, even if there is no solar radiation measurement equipment, because the main variable of these models being the temperature is one of the most available and easy to measure variables, having said this and after finding the mean square error, the mean absolute percentage error, the mean absolute bias error and the determination coefficient (R2) it was found that the most functional model and of better performance was model 6, this supported in a program developed in the programming language `C++`. [33].

$$\frac{H}{Ho} = a * T^{b} * Go + c \tag{28}$$

where a, b and c are empirical coefficients, T is the monthly average of temperature in degrees Celsius, and Go is the extraterrestrial sola radiation in MJ/m^2day

3.15. Models based in other meteorological parameters:
An average of data is required to establish an accurate estimate of the solar radiation value, however, these may be very scarce, so it has been decided to use meteorological parameters other than the classic ones, such as precipitation, relative humidity, dew point temperature, soil temperature, evaporation, pressure, wind speed and rainfall [17].

3.16. Sabbagh Model:
This model is proposed to be applied in dry arid or semi-arid regions such as Iran, which are located at an average sea level, i.e. approximately 300 m, and in case the area or region for which they plan to use this model has a higher altitude, it will be necessary to modify it, this model, although it could consider the effect of relative humidity, cannot take into account the effect of dust; This is how the geographic and seasonal factor called Kg [34].

$$H = 0.06407(Kg)exp\left[L\left(\frac{S}{12} - \frac{RH^{0.333}}{100} - \frac{1}{T_{max}}\right)\right] \quad (29)$$

where Kg is a geographical and seasonal factor, RH relativity humidity, and T_{max} it the maximum temperature

3.17. Sabziparvar Model:
This model modifies the Sabbagh model including three new considerations such as: correction factor for altitude, correction factor distance sun land and the monthly average of the total number of days with dust, in addition to this the fact of the variation of the hours of sun that would be assumed to be 12 for any place in the world, but for latitudes considered subtropical or that exceed or are equal to 30 degrees of latitude, the variation should be considered in a different way to avoid errors of overestimation or underestimation of solar energy. [34].

3.18. Maghrabi Model:
This model was created using 9 years of solar radiation data, it was established for Tabouk in Saudi Arabia (28.38-N, 36.6-E), it takes into account five meteorological parameters for its creation, which are: perceptible water vapor, temperature, relative humidity, atmospheric pressure and possible daily sunshine hours, this model has a high accuracy, to which different statistical tools were applied in its creation, such as mean percentage error (MPE), root mean square error (RMSE), mean bias error (MBE), and correlation coefficient (R); This model was compared with 29 other models available in the literature and showed an underestimation or overestimation of measured solar radiation, the model that showed a better prediction with respect to the model developed was the Abdalla model, this does not mean that the other models are not accurate, but that being created for different places in the world, they have different pre-established conditions that in this case do not make them the most suitable. This is applicable to any place that has conditions similar to those of Tabouk in Saudi Arabia. [35].

$$H = 163.01 - 1.04\left(\frac{S}{So}\right) + 0.12 * T - 0.21 * P - 1.06 * PWV - 0.03 * RH \quad (30)$$

where: T is the temperature in degrees Celsius, P is the atmospheric pressure, RH is the relativity humidity, and PWV is the Perceptible water vapor in mm.

As technology has advanced, use has been made of computational tools such as: Matlab which allows to simulate, model and determine characteristics, with the blocks already preset in Simulink [36], or with neural networks as for example in: Estimation of Hourly Solar Radiation Using Empirical Models and Artificial Neural Networks [37] where they use as input variables solar radiation estimates obtained from a simple mathematical model and different easily acquired climatic variables such as: temperature, pressure and humidity, where in addition a comparison was made that showed that these turn out to be adequate or you can also count on applications such as: Arduino, which allows to create prototypes that record the measurement of solar radiation [38], or Machine Learning which programs and adapts a type of algorithms according to the need that will allow the machine to make predictions or identify different patterns, as is the case of: Solar radiation forecasting model based on Machine Learning where the objective is to reduce uncertainty, grouping data that are sent to a predictor that estimates numerical data and when compared with measured data establishes a deviation that allows to know not only the uncertainty but also to estimate a long-term energy reserve, In addition to these tools, there are also statistical methods that allow calculating the error with a certain number of established data, which in turn allows making a comparison to know which would be the most optimal model depending on each situation [39].

4. Conclusions

Solar radiation is a fundamental parameter for everything, as time goes by in our attempt to reinvent and build the future in a better way, research has led us to understand that this is an even broader topic than one would think, knowledge about solar radiation is not only important for electrical engineering but it is a knowledge that can be applied to multiple fields such as agronomy, meteorology, geography, even dermatology, to mention a few examples, since it directly or indirectly affects many of the natural and artificial processes that occur on the planet. Measuring equipment is expensive so they can not be available to everyone, so you can resort to institutions in this case for Colombia to the IDEAM (Spanish Acronym of Meteorological Institute of Colombia) or agricultural associations which may have such instrumentation and can provide the corresponding data as is the case of Cenicafé (National Center for Coffee Research). Undoubtedly, one of the most used models, despite the fact that its measured data are not very easy to obtain, is the Ångström-Prescott model, the different variations have allowed it to be applied in many parts of the world, however, this is very important since, as far as possible, these variables should be adjusted locally, to reduce the margin of error and not overestimate or underestimate the data obtained. The technological advance has allowed the creation of multiple tools, which allows that nowadays the calculation of the equations is first of all easier and that if it is applied in the correct way, it does not have errors, that is why it is so important to know the theory.

Depending on the data you have, we recommend the following:

If you have a sunshine data, the Ångström Prescott model is undoubtedly the best option, if you do not have data to perform the regression that allows you to calculate a and b, we recommend models with pre-established variables with Page, Bahel, Gopinathan, it all depends on what other data you may have. If you have a temperature data, but you do not have the value of the constant you can use the Hargreaves-Samani model or if you have the pressure you can use the Allen model, or if you have altitude you can use the Annandale model. If you have the relative humidity, temperature and latitude, which are easily accessible data, you can make use of the Sabbagh model, but if in addition to this you have the possibility of obtaining the atmospheric pressure the model you can use is Maghrabi.

Acknowledgments

Gloria Marina Bautista thanks to her family, specially her brother for his dedication, her mother for her company and her father for his support, to my children for their infinite love and unwavering strength to continue, to her friends for listening to me talk about the subject even without understanding and even to raise solutions to my doubts, to Diego thanks for helping me with Latex, without a doubt thanks to each and every one of the people who have contributed to my professional and personal growth,

I will be eternally grateful to them. We give thanks to Universidad Distrital Francisco José de Caldas and Electrical Engineering program by Cycles, CEM (Electromagnetic Compatibility) research group.

References

[1] Zerrahn A, Schill W P and Kemfert C 2018 *European Economic Review* **108** 259–279 ISSN 0014-2921 URL `http://dx.doi.org/10.1016/j.euroecorev.2018.07.004`
[2] Walston L J, Rollins K E, LaGory K E, Smith K P and Meyers S A 2016 *Renewable Energy* **92** 405–414 ISSN 0960-1481 URL `https://www.sciencedirect.com/science/article/pii/S0960148116301422`
[3] Garcia N A, Urrea D A O, Forero A M S, Castro G S, Torres D A, Medina J A B, Luengas O L R, Diaz S S and Alfonso O H R 2020 Zonas no interconectadas – ZNI Informe sectorial de la prestación del servicio de energía eléctrica
[4] García Gamba W A and Ovalle Sandoval L V 2017 Modelo para Determinar la Irradiancia Solar en Diferentes Zonas del Territorio Colombiano a partir de Imágenes Aéreas o Satelitales. accepted: 2017-06-13T17:11:57Z URL `http://repository.udistrital.edu.co/handle/11349/5714`
[5] Gustavsson M 2007 *Energy Policy* **35** 1292–1299 ISSN 0301-4215 URL `https://www.sciencedirect.com/science/article/pii/S0301421506001492`
[6] Juzeniene A, Brekke P, Dahlback A, Andersson-Engels S, Reichrath J, Moan K, Holick M F, Grant W B and Moan J 2011 *Reports on Progress in Physics* **74** 066701 ISSN 0034-4885 publisher: IOP Publishing URL `https://doi.org/10.1088/0034-4885/74/6/066701`
[7] Cabezas M 2005 *Agronomía Colombiana* **23** 12
[8] Guzman M O, Baldion R J V, Simbaqueva F O, Zapata H J and Chacón C C 2013 *Revista del Centro de Investigaciones de Café - CENICAFE* ISSN 0120-0275 URL `https://biblioteca.cenicafe.org/handle/10778/526`
[9] Florez C, Rodriguez A, Zapata H, Concha I, Romero L, Vesga D, Costa C, Henriquez M, Simbaqueva O, González O, Chacón C and Pérez J 2005 *Atlas de radiación solar de Colombia* (UPME-IDEAM) ISBN 958-8067-13-8 URL `http://bdigital.upme.gov.co/jspui/handle/001/1236`
[10] García R D, Cuevas E, García O E, Cachorro V E, Pallé P, Bustos J J, Romero-Campos P M and de Frutos A M 2014 *Atmospheric Measurement Techniques* **7** 3139–3150 URL `https://amt.copernicus.org/articles/7/3139/2014/`
[11] Rodríguez-Patarroyo D J, Hernandez-Mora J and Simbaqueva-Fonseca O 2015 *Revista Científica* **21** 65–70 URL `https://revistas.udistrital.edu.co/index.php/revcie/article/view/8457`
[12] Simbaqueva L and Suescún N 2001
[13] Crabtree G W and Lewis N S 2007 *Physics Today* **60** 37–42 includes discussion of solar thermoelectric conversion URL `http://link.aip.org/link/?PTO/60/37/1`
[14] Iqbal M 1983 *An introduction to solar radiation* (Toronto ; New York: Academic Press) ISBN 978-0-12-373750-2 978-0-12-373752-6
[15] Reyes F D 2013 *Modelo numérico para la generación de mapas de radiación solar y su aplicación al aprovechamiento de energía solar fotovoltaica y termoeléctrica* http://purl.org/dc/dcmitype/Text Universidad de Las Palmas de Gran Canaria URL `https://dialnet.unirioja.es/servlet/tesis?codigo=86117`
[16] Angstrom A 1924 *Quarterly Journal of the Royal Meteorological Society* **50** 121–126
[17] Besharat F, Dehghan A A and Faghih A R 2013 *Renewable and Sustainable Energy Reviews* **21** 798–821 ISSN 1364-0321 URL `https://www.sciencedirect.com/science/article/pii/S1364032112007484`
[18] Al-Mostafa Z A, Maghrabi A H and Al-Shehri S M 2014 *Energy Conversion and Management* **84** 209–216 ISSN 0196-8904 URL `https://www.sciencedirect.com/science/article/pii/S0196890414003252`
[19] Poveda Matallana W D 2020 Validación de la radiación solar en superficie para la región Orinoquía a partir de imágenes de satélite accepted: 2020-08-08T16:37:36Z Publisher: Bogotá - Ciencias - Maestría en Ciencias - Meteorología URL `https://repositorio.unal.edu.co/handle/unal/77981`
[20] IDEAM-UPME 2015 Atlas de radiación solar, ultravioleta y ozono de Colombia URL `http://atlas.ideam.gov.co/presentation.`
[21] Glover J and McCulloch J S G 1958 *Quarterly Journal of the Royal Meteorological Society* **84** 172–175 ISSN 1477-870X _eprint: https://onlinelibrary.wiley.com/doi/pdf/10.1002/qj.49708436011 URL `https://onlinelibrary.wiley.com/doi/abs/10.1002/qj.49708436011`
[22] Page J K 1961 The estimation of monthly mean values of daily total short wave radiation on vertical and inclined surfaces from sunshine records for latitudes 40N-40S by / (UN,) URL `https://digitallibrary.un.org/record/3827996`

[23] Bahel V, Bakhsh H and Srinivasan R 1987 *Energy* **12** 131–135 ISSN 0360-5442 URL https://www.sciencedirect.com/science/article/pii/0360544287901174
[24] Benson R B, Paris M V, Sherry J E and Justus C G 1984 *Solar Energy* **32** 523–535 ISSN 0038-092X URL https://www.sciencedirect.com/science/article/pii/0038092X84902676
[25] Gopinathan K K 1988 *Solar Energy* **41** 499–502 ISSN 0038-092X URL https://www.sciencedirect.com/science/article/pii/0038092X88900527
[26] Zabara K 1986 *Solar & Wind Technology* **3** 267–272 ISSN 0741-983X URL https://www.sciencedirect.com/science/article/pii/0741983X86900056
[27] Ajayi O O, Ohijeagbon O D, Nwadialo C E and Olasope O 2014 *Sustainable Energy Technologies and Assessments* **5** 28–36 ISSN 2213-1388 URL https://www.sciencedirect.com/science/article/pii/S2213138813000702
[28] Bristow K L and Campbell G S 1984 *Agricultural and Forest Meteorology* **31** 159–166 ISSN 0168-1923 URL https://www.sciencedirect.com/science/article/pii/0168192384900170
[29] Hargreaves G H and Samani Z A 1982 *Journal of the Irrigation and Drainage Division* **108** 225–230 publisher: American Society of Civil Engineers URL https://ascelibrary.org/doi/abs/10.1061/JRCEA4.0001390
[30] Hunt L A, Kuchar L and Swanton C J 1998 *Agricultural and Forest Meteorology* **91** 293–300 ISSN 0168-1923 URL https://www.sciencedirect.com/science/article/pii/S0168192398000550
[31] Annandale J, Jovanovic N, Benadé N and Allen R 2002 *Irrigation Science - IRRIG SCI* **21** 57–67
[32] Allen R G 1997 *Journal of Hydrologic Engineering* **2** 56–67 ISSN 1084-0699 publisher: American Society of Civil Engineers
[33] Hassan G E, Youssef M E, Mohamed Z E, Ali M A and Hanafy A A 2016 *Applied Energy* **179** 437–450 ISSN 0306-2619 URL https://www.sciencedirect.com/science/article/pii/S0306261916309473
[34] Sabziparvar A A 2008 *Renewable Energy* **33** 1002–1010 ISSN 0960-1481 URL https://www.sciencedirect.com/science/article/pii/S0960148107002248
[35] Maghrabi A H 2009 *Energy Conversion and Management* **50** 2754–2760 ISSN 0196-8904 URL https://www.sciencedirect.com/science/article/pii/S0196890409002477
[36] Shah M W and Biate R L 2016 *International Journal of Scientific & Engineering Research* **7** 551–554
[37] Jimenez V A, Will A and Rodríguez S 2017 *Ciencia y tecnología* 29–45 ISSN 1850-0870, 2344-9217 publisher: Universidad de Palermo (UP) Section: Ciencia y tecnología URL https://dialnet.unirioja.es/servlet/articulo?codigo=6223470
[38] Alves P V, Reis L H S, Querino C A S, Moura M A L, Feitosa Júnior A A and Martins P A d S 2020 *Revista Brasileira de Ensino de Física* **42** ISSN 1806-1117, 1806-9126 publisher: Sociedade Brasileira de Física URL http://www.scielo.br/j/rbef/a/zZZgcdNLCqvXzQY736GbyTj/abstract/?lang=en
[39] Khatib T, Mohamed A and Sopian K 2012 *Renewable and Sustainable Energy Reviews* **16** 2864–2869 ISSN 1364-0321 URL https://www.sciencedirect.com/science/article/pii/S1364032112000767

COMSOL modeling of the lightning-generated electric field distribution in a chain of insulators of a 115 kV power transmission line.

A F Cadena[1], J S Mendoza[2], and H F Ibáñez[3]

Universidad Distrital Francisco José de Caldas, Facultad Tecnológica, Cra. 7 # 40B-53, Bogotá, Colombia

E-mail: [1]afcadenaf@correo.udistrital.edu.co, [2]jsmendozag@correo.udistrital.edu.co, [3]hfibanezo@udistrital.edu.co

Abstract. The transmission lines installed in our city, to the outskirts of it and in the rest of the national territory produce a level of emissions of electric field, these levels vary depending on the physical disposition of the drivers, the distance between spans, the voltage level of the line, among others. These values must respect the exposure limits for individuals, as stipulated in Article 14 of RETIE. In this work will be developed a study of distribution of the electric field along a chain of insulators standard of tempered glass of a transmission line of 115 kV with the aid of the software COMSOL Multiphysics, to the impact of an atmospheric discharge on the storage cable with the help of ATPDraw software, taking into account variations in the value of the earthing resistance of the structure. With the electric field distribution values obtained by means of simulations carried out in the COMSOL Multiphysics software, comparisons are made of the magnitude of the obtained field, taking as a reference the stable state of the system without condition of pollutants. Initially a simulation of a transmission line of 115 kV is performed in the ATPDraw software, analyzing the lightning impulse generated on the guard cable, at different values of grounding resistance, to determine the waveform and reached value of overvoltage in the insulator chain. Then the overvoltage signal obtained with the help of ATPDraw software is recreated in the COMSOL Multiphysics software, to be able to visualize the distribution and behavior of the electric field along the chain, taking into account the different factors involved in the process, pollutants such as salinity, themes such as travelling waves etc. and thus determine possible line isolation failures by comparing the values achieved by simulating in COMSOL Multiphysics with respect to the CFO of the insulator chain.

1. Introduction

Colombia has a high density of lightning to ground because it is located in the Intertropical Confluence Zone, so it has one of the largest lightning activities on the planet. Therefore, it is essential for the design and commissioning of a power transmission line, to consider the electrical and constructive characteristics of the line and the location where it is installed. These parameters are necessary to comply with the number of outputs required by the network operator or failing that, which is established in resolution CREG 025 of 1995, which establishes that for voltage levels equal to or greater than 220 kV, they are 3 outputs per 100 km of line per year. For this reason, by means of an isolation coordination, it must be demonstrated that the number of outputs calculated is below the established value.

For a network of 115 kV (study voltage level) there is no regulation around the requirement of a certain number of outputs of the lines by lightning, however, the resistivity of the soil of the installation of the line and the grounding resistance must be considered. For example, in the department of Meta between the municipalities of Villavicencio and Guamal, the company Electrificadora del Meta S.A. E.S.P (EMSA) for the transmission line between the Ocoa and Guamal substations proposed a maximum value of 10 outputs per 100 km of line per year, since they take into account the high resistivities of the installation grounds of the line, because the greater the grounding resistance, exist a high probability that flash over will occur and therefore greater is the number of outputs that can occur [1]. One of the anomalies within an electrical power system are the overvoltages due to lightning that cause a greater electrical stress in the transmission systems, causing the electrical damages due to the natural phenomenon, since the transmission lines cover a very large area within the land where they are installed, given that their sections cover long distances and cross lands exposed to different types of climates.

Therefore, the present paper models with the help of Comsol and ATP software the electric field distribution in a chain of insulators of a 115 kV transmission line with contamination, to visualize the electric field values that occur when the network is in normal operation and when lightning strikes the structure.

2. Methods

2.1. ATPDraw modeling of a transmission line in the event of an atmospheric discharge.

For the present study, the results of the fast front overvoltages in a typical 115 kV line are presented, in order to obtain the most representative overvoltage values in the event of an atmospheric discharge in the area. Therefore, the above is considered from the NTC 4552-1 standard, taking into account the conditions under which the line operates. In addition, the IEC 60815-2 standard is taken into account in order to consider the parameters of the area in which the transmission line is operating. The peak lightning current magnitude is 43 kA and its representative current shape is the standard atmospheric impulse (10 / 350 μs). This allows to obtain the waveform of the overvoltage originated at the point of impact of the transmission tower.

It is essential for any type of simulation to select the appropriate models for the lightning surge analysis scenario. For the modeling of each of the elements of the system, the standard IEC 60071-4 [2], [3], Standard NTC 4552-1 [4] and [5] have been taken into account.

In the particular case of Colombia, the measurement data correspond to studies carried out with different measurement methods. The lightning current amplitude was estimated by means of vertical electric field measurements carried out in 1995 by means of a parallel plate antenna, previously calibrated in the laboratory using a high resolution digital oscilloscope and associated measuring equipment. These data were compared with those recorded by the TSS-420 storm sensor in operation at the National University facilities in Bogota. The impact distance was calculated using information provided by the Colombian lightning location network RECMA.

Figure 1 shows the comparative probability results between the values given by CIGRE in 1979 for records taken in non-tropical latitudes and those estimated in four tropical countries: Brazil (Cachimbo station, State of Minas Gerais, 1996), Rhodesia (Anderson, et. al., 1954), Malaysia (Lee, et. al., 1979) and Colombia (Torres, et. al., 1995). This graph shows the higher probability of negative lightning return current magnitude in tropical zones (Brazil, Malaysia, Colombia and Rhodesia), with respect to non-tropical zones (CIGRE) [4].

Journal of Physics: Conference Series **2135** (2021) 012006 doi:10.1088/1742-6596/2135/1/012006

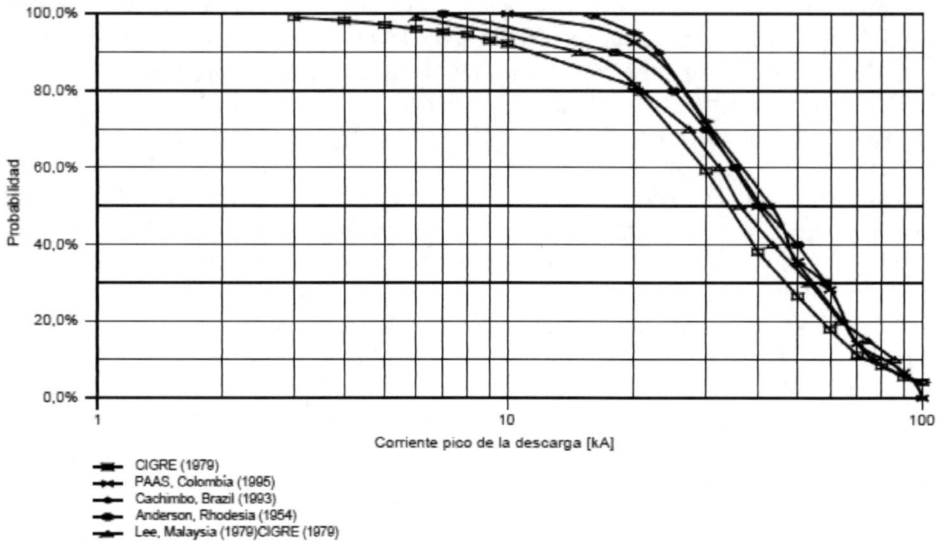

Figure 1. Cumulative probability curve for negative return current [4].

The surge impedance of the towers of a transmission line varies along the tower and is affected by the wave propagation time, so it depends on the details of the structure [6]. Therefore, the tower can be represented as a single-phase transmission line with an impulse impedance and a wave propagation speed equal to that of light [6].

For power systems, the model of atmospheric impulses must take into account that the reflections of the traveling wave due to the impact of lightning must not return to the analysis node in order to avoid adding to the wave that arrives at that point. This is achieved by modeling a line of infinite length, i.e., a line impedance that avoids reflections of the traveling wave on the transmission line [7].

To determine the length of the chain or isolation distance, the IEC 60815-2 and IEC 60071-2 standards must be considered. Therefore, the zone in which the transmission line is located must be determined in order to obtain the RUSCD, which according to IEC 60815-2 is defined as the normal unified specific leakage distance for each level of contamination. Figure 2 shows the RUSCD or USCD value depending on the level of contamination present in the study area.

For standardization purposes, five classes of contamination characterizing the site are defined in IEC 60815-2 [8], from very light contamination, to very heavy contamination as:

a. Very light
b. Light
c. Medium
d. Heavy
e. Very heavy

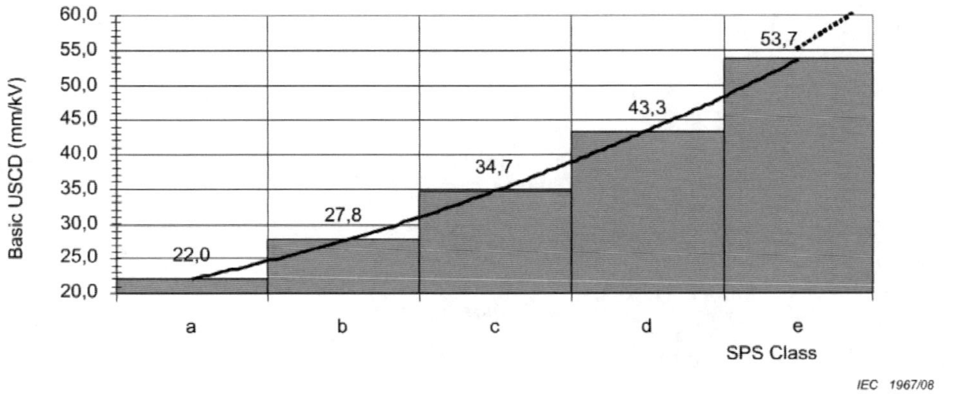

Figure 2. RUSCD as a function of the severity of the contamination class in the area [8]

The procedure for determining the isolation distance is based on equation (1).

$$USCD_C = RUSCD * K_a * K_{ad} \quad (1)$$

Where:

$USCD_C$ = Corrected unified specific leakage distance.
$RUSCD$ = Nominal unified specific creepage distance
K_a = Altitude correction factor
K_{ad} = Correction factor according to insulator diameter

2.2. Transmission line modeling.

For the modeling of phase conductors against atmospheric impulses for transmission lines, the Bergeron model of the LINE CONSTANTS subroutine of the ATP was considered, taking into account that it is a model of conductors with constant distributed parameters, which is based on wave propagation in a defined section. The frequency used for the model is 100 kHz [7]. In order to know the simulated overvoltage values at any point of the transmission line, the input values are defined for the ATP (Alternative Transient Program) software - Version 6.0.

The electrical parameters of the system under study are

- Rated voltage: 115 kV
- Maximum service voltage Um: 123 kV
- Typical line span: 300 m

The parameters of the phase and guard conductors are presented in Tables 1 and 2.

Table 1. Phase conductor parameters [a].

Conductor HAWK ACSR 477 kcmil (typical conductor for this voltage level)	
Aluminum wire diameter	3.439 mm
Galvanized steel wire diameter	2.674 mm
Conductor outside diameter	21.79 mm
Steel section	39.31 mm²
Aluminum Section	241.5 mm²
Total Section	281 mm²
DC resistance of the conductor at 25 °C	0.1198 Ohm/km
AC conductor resistance at 75 °C	0.1432 Ohm/km

[a] Values taken from the manufacturer's catalog CENTELSA.

Tabla 2. Guard conductor parameters [a].

OPGW 24 SM-14.4 mm ACS de 24 fibers (typical conductor for this voltage level)	
Conductor	OPGW
Type	ACS (Aluminium Clad Steel)
Caliber	24SM-14.4 mm
Outside diameter	14.4 mm
DC resistance	0.427 Ohm/km
OPTICAL UNIT	
Number of Fibers	24
Outer Diameter Fiber Container Tube	3.2 mm
DC resistance	3.76 Ohm/km

[a] Datasheet cable OPGW 24 SM-14.4 mm

The layout of the phase conductors and guard wires in the structure for a 115 kV network is shown in Figure 3, which corresponds to a double-circuit structure in vertical configuration.

2.3. *Lightning impulse source.*

For the model of the atmospheric discharge that will impact the transmission tower under study, the ATP Heidler type source is used, with a peak current of 43 kA, a wave rise time of 10 μs and a decay time of 350 μs. The selected amplitude value was extracted from the NTC 4552-1 standard (Protection against atmospheric electric discharges (Lightning) Part 1: General principles - Table A.1) Medians of the peak value of the lightning return current in different areas of the planet [4].

2.4. *Chain of insulators.*

For the modeling of the insulator strings, voltage controlled switches were used, as proposed in [7], taking into account that the Critical Flash Overvoltage (CFO) for each of the insulator strings corresponds as shown in equation (2). Figure 4 shows the element used SwitchVC - Switch Voltage Control of ATPDraw.

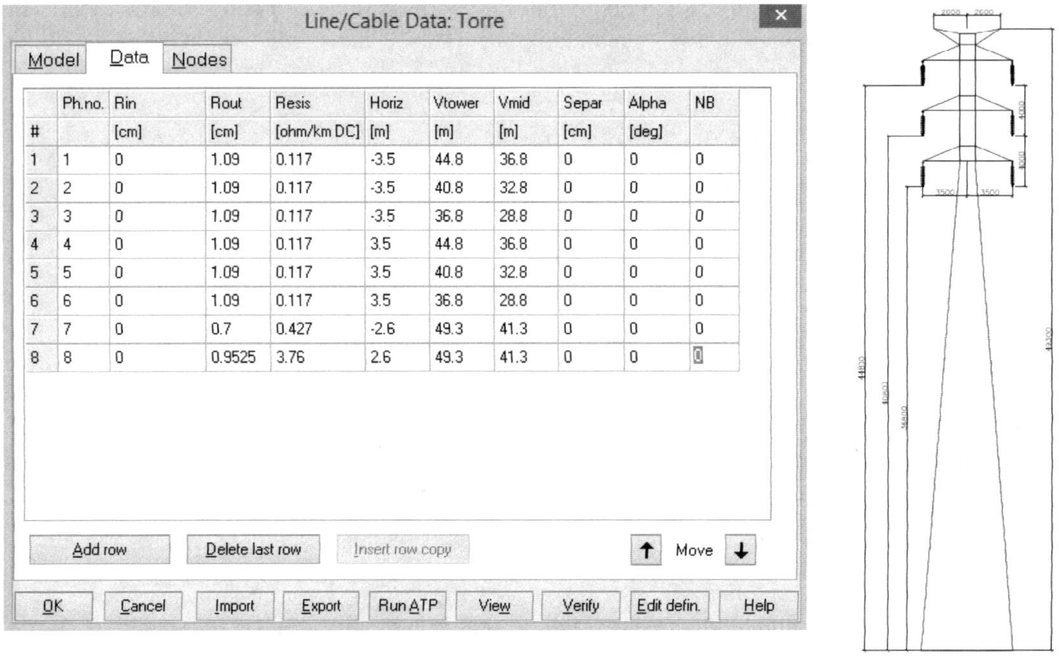

#	Ph.no.	Rin [cm]	Rout [cm]	Resis [ohm/km DC]	Horiz [m]	Vtower [m]	Vmid [m]	Separ [cm]	Alpha [deg]	NB
1	1	0	1.09	0.117	-3.5	44.8	36.8	0	0	0
2	2	0	1.09	0.117	-3.5	40.8	32.8	0	0	0
3	3	0	1.09	0.117	-3.5	36.8	28.8	0	0	0
4	4	0	1.09	0.117	3.5	44.8	36.8	0	0	0
5	5	0	1.09	0.117	3.5	40.8	32.8	0	0	0
6	6	0	1.09	0.117	3.5	36.8	28.8	0	0	0
7	7	0	0.7	0.427	-2.6	49.3	41.3	0	0	0
8	8	0	0.9525	3.76	2.6	49.3	41.3	0	0	0

Figure 3. Arrangement and distances of phase and guard wire conductors in vertical configuration for typical 115 kV structures.

$$CFO = \left(400 + \frac{710}{t^{0.75}}\right) x\, W \quad [kV] \tag{2}$$

Where:
t: 6 µs flashover time.
W: Length of insulator chain.

For this article a study area is established, corresponding to the city of Bogotá located in the department of Cundinamarca, in such a way that taking a Very Heavy contamination level, the number of isolators (NA) is calculated, giving the results shown in Table 3.

Tabla 3. Contamination level and resulting isolation distance for each case of study

Case study	Altitude above sea level [m.a.s.l.].	Contamination level	NA (calculated)	NA (standardized)
Bogotá D.C	2600	Very Heavy	16,4	17

Once the total length of the insulator chain has been calculated, the CFO (Critical flash Overvoltage) is calculated for each case of study according to equation (2). Thus, this value for the network in the city of Bogotá is 1452 kV.

2.5. Transmission tower structures.
This model takes into account what Sargent and Darveniza proposed for this type of structures, which corresponds to a conical model providing a constant impedance for the transmission tower [9].

- Conical Model

$$Z_t = 60 \ln\left(\sqrt{2}\frac{\sqrt{r^2 + h^2}}{r}\right) \tag{3}$$

Where:
Z_t = Surge impedance of the structure.
h = Tower height (m).
r = Tower base radius (m).
c = Speed of light($3x10^8\ m/s$).

The structure data are shown in Figure 3 to obtain the surge impedance values of the structure for each modeled part. Considering the above, a distributed parameter model of a Clarke-type single-phase line was implemented to represent the transmission structures, considering a typical grounding resistance value for 115 kV structures of 20 Ω. For the modeling of each of the associated structures, a LINEZ characteristic impedance segment used by the ATP software was considered.

3. Data export from ATP to Comsol software.

Figure 4 shows a section of the modeled system and the site of the impact of the atmospheric electric discharge and where the electric overvoltage presented in the chain of insulators of the transmission line is evaluated. Figure 5 presents the result of the overvoltage obtained with the help of the ATP software, considering a grounding resistance of 20 Ω. This signal obtained and shown in Figure 5 is exported to the Comsol software to visualize the electric field distribution along the insulator chain.

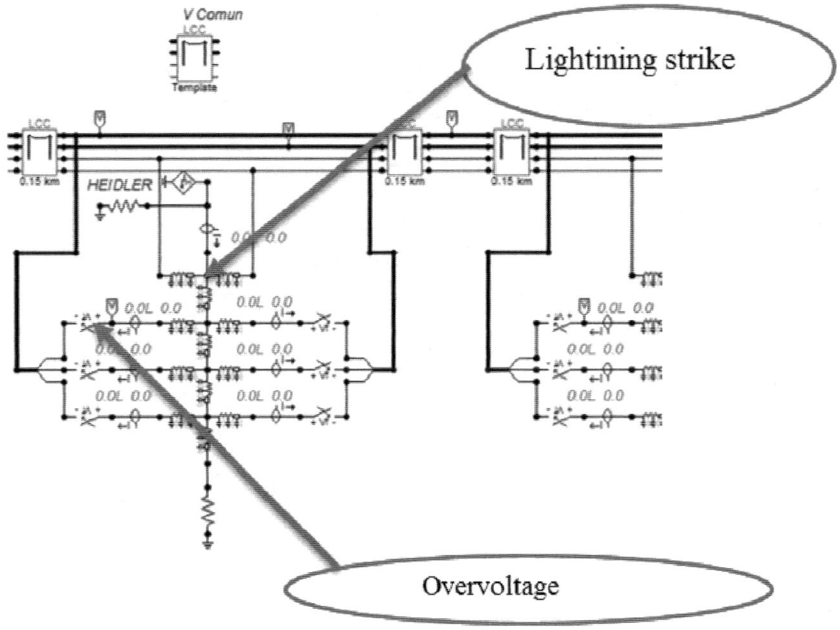

Figure 4. Measurement terminal, overvoltage signal tower adjacent to the impacted tower.

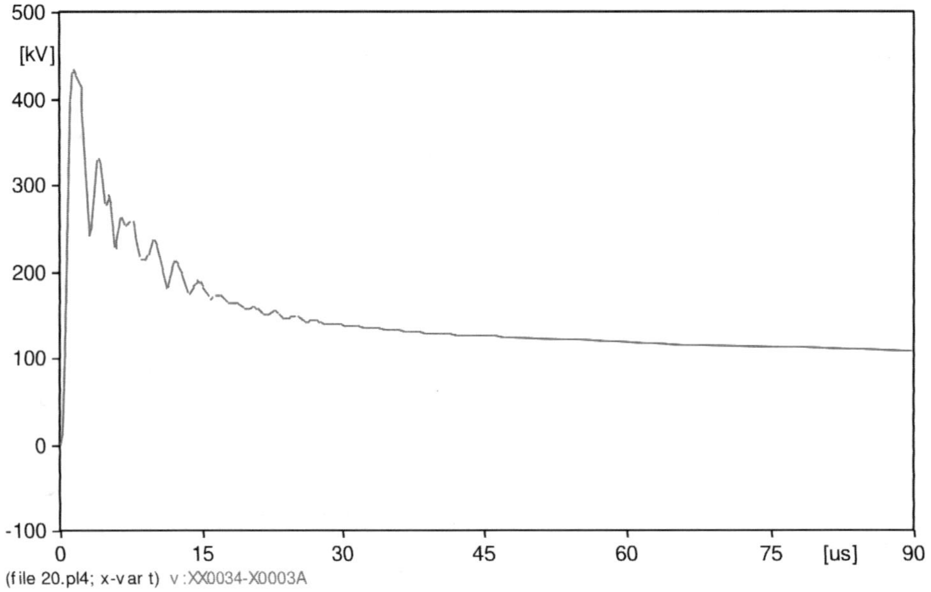

Figure 5. Overvoltage due to atmospheric discharge with 20 Ω grounding resistance.

4. Results Potential and electric field distribution along the insulator chain of the 115 kV line subjected to atmospheric discharge and contamination.

Figure 6 shows the case study where an atmospheric discharge impacts the transmission tower and the overvoltage induced in the terminal corresponding to the grounded part of the transmission tower is obtained and imported into the Comsol software. In this way, the resulting electric field is visualized in that area at the time of the highest overvoltage present and to determine if there is a break in the dielectric rigidity of the air.

This overvoltage presents a peak value of 435 kV in the fitting that allows fixing the insulator chain in the crossarm, which allows defining the distribution of the potential along the chain. As we know, this overvoltage does not interfere in the normal operation of the system, since it is within the range of support of the insulator chain, better known as the CFO. However, we seek to study and analyze the distribution of the potential and electric field in this small instant of time where this overvoltage is generated.

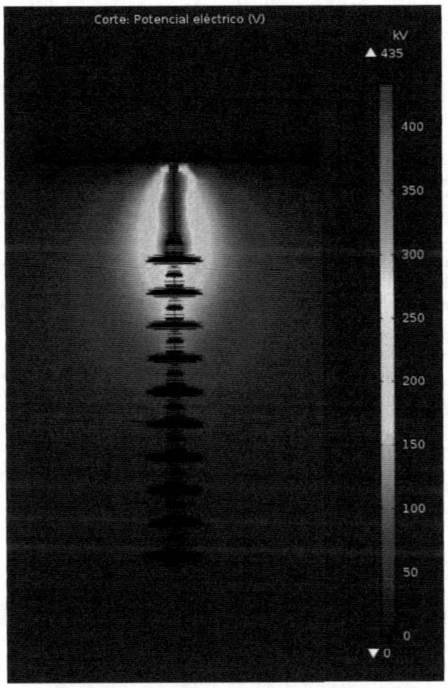

Figure 6. Potential distribution in a chain of insulators in the event of an atmospheric discharge on a crossarm or structure.

The potential distribution is studied with the first peak of the imported ATP wave, corresponding to a time t = 1.36 μs as the most critical case. Figure 7 shows the behavior of the electric potential along the insulator chain.

Figure 7 shows a peak value of 435 kV up to an arc length of 354 mm where the length of the fitting plus the skull attached to it is positioned. Subsequently, the potential of the other insulator units can be detailed from top to bottom according to figure 6 when it behaves as a constant, it represents the section comprised by the junction of the bolt of the first insulator and the skull of the second insulator. Finally, the interaction of the cement and the glass in each insulating unit is observed, causing small protuberances to appear on the curve, thus showing how the potential tends to decrease at that instant of time.

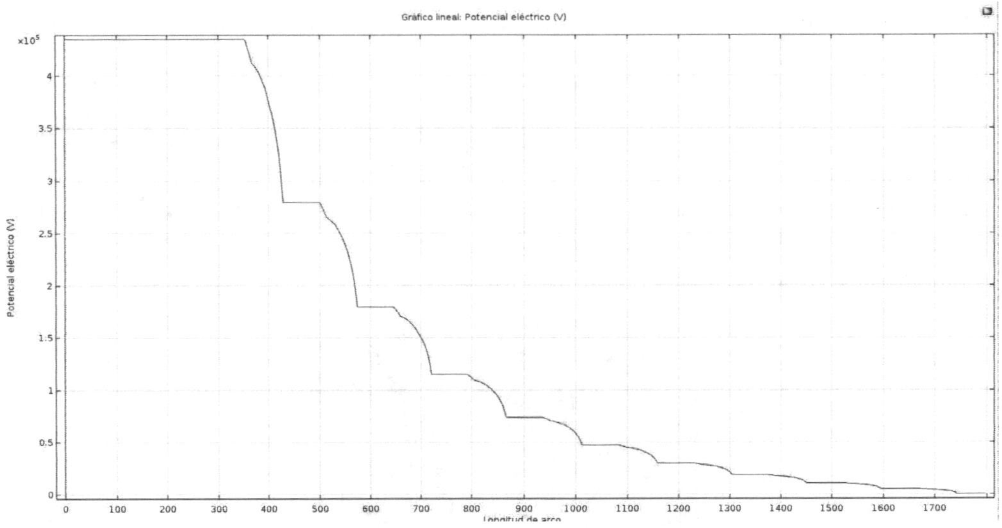

Figure 7. Potential distribution in a chain of uncontaminated insulators in the event of an atmospheric discharge in a crosshead or structure by means of a linear graph.

The results obtained show that the concentration of pollution in the surface layer of the insulator string does not impact the results of the electric potential along the string, i.e. there is no variation in the increase of the potential even when it is affected by an atmospheric discharge.

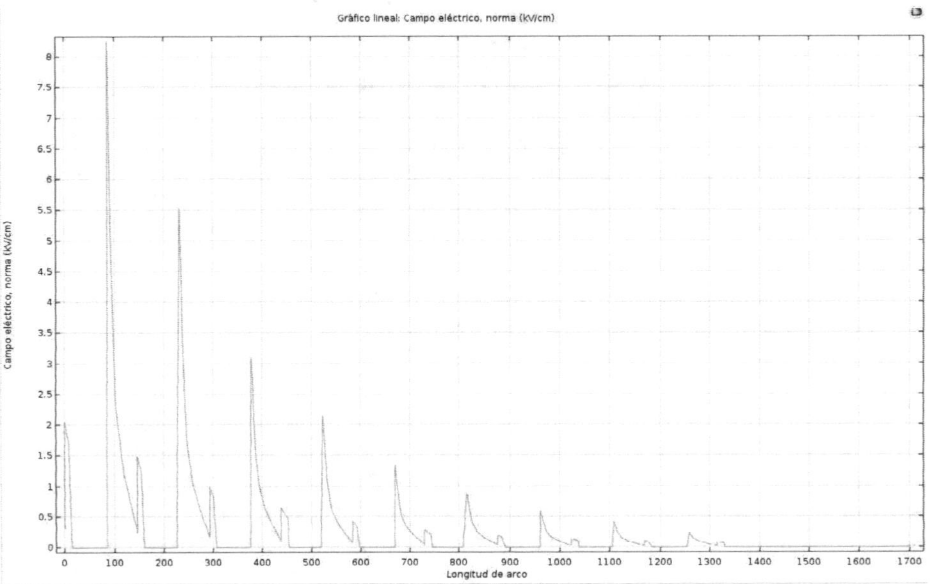

Figure 8. Line graph: distribution of the electric field standard (kV/cm) along the insulator chain, under normal operating conditions and without contamination.

The distribution of the electric field in the center of the chain of insulators operating normally and without contamination is shown in Figure 8, which highlights a higher level of electric field magnitude in the energized part (115 kV line), and as the insulation level is increased approaching the equipotential connection of the tower, the level of the field magnitude decreases reaching a value of 0 kV/cm. It can be observed that the distribution of the electric field magnitude in the first 3 insulators is reduced from

Journal of Physics: Conference Series **2135** (2021) 012006 doi:10.1088/1742-6596/2135/1/012006

8.2 kV/cm to 3 kV/cm (electric field in the third insulator), approximately, as shown in Figure 8. This means a decrease of 63%, so that for the subsequent 7 insulators of the chain the rate of change is lower, proving the non-linear behavior of the electric field magnitude in the chain.

The distribution of the electric field along the chain considering the contamination and the incidence of an atmospheric discharge is shown in Figure 9. An increase in the electric field associated with the first peak is influenced by the contamination in the area where the transmission line operates, which has an electric field value of 72 kV/cm in the second peak, an increase of about 1.4 times the value of the electric field if the contamination had not been taken into account.

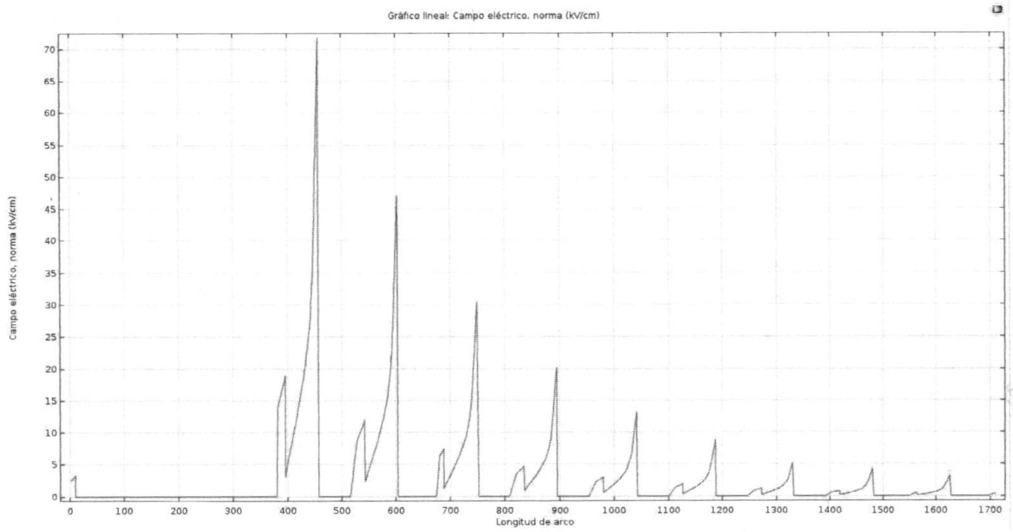

Figure 9. Electric field distribution along the chain of insulators with contamination in the presence of an atmospheric discharge in a crosshead or structure by means of a line graph.

Likewise, the first peak, as indicated in the previous case, presents a value of 18.8 kV/cm between the cement that joins the skull and the glass, followed by the second peak, which increases from 51 kV/cm in the case where contamination is not taken into account to 72 kV/cm taking into account contamination. It can be observed in the study that in the case where no contamination is taken into account, the electric field in the last insulating unit is 3 kV/cm at an arc length of approximately 1595 mm, in contrast to the electric field value in the same insulating unit of 3.2 kV/cm according to Figure 9, but at an arc length of approximately 1625 mm, which evidently shows the indicated increase.

5. Conclusions.

The insulators when subjected to contamination conditions in the environment generate accumulation on its surface, which the behavior of the electric potential distribution is not subject to any variation, on the other hand, it generates interaction of the electric field with the accumulation of salinity on the surface, consequently leading to the increase of the same similar to what was found by [10], which can generate with certain particular conditions of the environment (such as those specified in this document), the generation of partial discharges on the surface of the insulator known as Flashover.

Regarding the analysis of the electric field implemented to the three-dimensional model, performed by the COMSOL Multiphysics software through the FEM, when designing a detailed geometry and being a robust analysis system, this requires computer equipment with high levels of data processing, since without this the model would take too much time in the process to obtain results or the possible nonconvergence of the model. Therefore, it is advisable to design processes in which the analysis is performed on simplified geometries or the development of dimensional models.

Is evident that when an atmospheric discharge occurs in the transmission line, the network is not always out of service, because the level of support to electrical stresses in the chain of insulators allows the system to withstand these eventualities since an adequate coordination of insulation for the line has been previously carried out.

In the evaluation of the dielectric stress of the air on one side of the insulator chain, it is observed that it does not exceed the value of the dielectric strength of the same in its total length, it is possible to think that in the first three insulators (from top to bottom) air rupture occurs, which would lead to a possible failure in the insulator chain. Therefore, thanks to the total length of the chain, the other insulators do not allow the arc to be generated between the phase-ground terminals.

The results obtained from the electric field under contamination conditions and when subjected to an atmospheric discharge, it was evidenced that the field value rises considerably on the surface of the insulator due to the conductivity of the contamination in the area, this contaminant layer damages the whole insulator. It was observed that the values of the electric field increase as it approaches the energized terminal, so that a bad maintenance to the line could cause along the chain of insulators an electrical disruption in the air and consequently an inverse flame.

Acknowledgments
The authors wish to acknowledge and thank the Technological Faculty of the Universidad Distrital Francisco José de Caldas for the support and funding during the course of this research work.

References

[1] Empresa Electrificadora del Meta, Normas de Construcción, Electrificadora del Meta S.A. E.S.P., Villavicencio, 2013.

[2] I. E. Commission, IEC 60071-4. Insulation co-ordination - Part 4: Computational guide to insulation co-ordination and modelling of electrical networks, IEC, 2004.

[3] L. H. Restrepo, G. Caicedo and F. Castro-Aranda, Transmission line models for electromagnetic, *Ingeniería y Computación,* vol. 16, no. 1, pp. 21-32, 4 Julio 2008.

[4] ICONTEC, Colombian Technical Standard 4552-1: Protection against lightning. Part 1: General principies, ICONTEC, Bogotá D.C., 2008.

[5] J. A. Martínez Velasco, Insulation coordination in high voltage networks, España: McGraw-Hill, 2007.

[6] J. Martínez Velasco and F. Castro Aranda , Modeling Of Overhead Transmission Lines For Lightning Overvoltage Calculations, *Revista Chilena de Ingeniería* , vol. 18, no. 1, pp. 120-131, 15 marzo 2010.

[7] L. Cardona, Medellín: Universiad Nacional de Colombia, 2004.

[8] I. E. Commission, Selection and dimensioning of high-voltage insulators intended for use in polluted conditions - Part 2: Ceramic and glass insulators for a.c. systems, IEC, London, 2008.

[9] G. E. Company, California: Electric Power Research Institute, 1982.

[10] C. Domínguez López, Modeling in ATPDRAW of the lightning behavior of a transmission line, Universidad Central Marta Abreu de las Villas, Santa Clara, 2004.

Equation of lightning return current in tropical zones

J Rubiano-Cruz[1], **S Páez-Moncaleano**[2], , **C Chacón-Cardona**[3], **and H Torres-Sánchez**[4]

[1]Profesor, Ingeniería Eléctrica, Facultad Tecnológica, Universidad Distrital Francisco José de Caldas, Bogotá D.C., Colombia
[2]Profesor, Departamento de estadística, Universidad Nacional, Bogotá, Colombia
[3]Profesor, Ingeniería Eléctrica, Facultad Tecnológica, Universidad Distrital Francisco José de Caldas, Bogotá D.C., Colombia
[4]Professor Emérito, Ingeniería Eléctrica, Universidad Nacional, Bogotá D.C., Colombia

E-mail: jjrubianoc@correo.udistrital.edu.co

Abstract. The purpose of this study is to model mathematically the equation of the lightning return current in tropical zones and to validate it with data obtained from direct measurements in Colombia and Brazil. This study is carried out because the lightning return current in tropical zones is greater than in other latitudes of the planet. For this reason, it is very important for different applications in electrical engineering. Nowadays, the Heidler equation is used for the simulation of faults in electrical systems. This equation will serve as a reference to make a comparison between the data of the equation obtained in this article with those from the equation of Heidler.

1. Introduction

Colombia and Brazil are countries located in the tropical zone of the planet[1] This zone has an elevated ground flash density and a lightning peak current higher than in other latitudes[2]. This atmospheric condition in tropical zones can induce potentially elevated problems to people and electrical systems.

Colombia has a high activity by lightning because of the aforementioned. For this reason, this phenomenon is studied in the country, since approximately 67 people die by lightning in the country per year, according to data of the Departamento Administrativo Nacional de Estadística (DANE) or in English National Administrative Department of Statistics, the entity in charge of the statistics of the country[3]. In terms of faults in the Colombian electrical system, the most common damages in distribution transformers is by an electric atmospheric discharge[4], due to lightning. Thus, it is important to study and search for alternatives to lower the indices of deaths and damages in electrical systems in the country.

Colombia and Brazil have direct lightning measurement stations because of their high activity by lightning, in which direct lightning signals have been recorded and measured. Therefore, this article proposes to develop and verify a mathematical model of the lightning return current for tropical zones. In addition, the data obtained from the model of this article will be compared with the equation of Professor Heidler[5], which is currently used for engineering designs and software simulations such as EMTP-ATP, that brings the Heidler type font in its library of elements.

2. Lightning peak current in the world

The first measurements of the lightning peak current were made in temperate zones. The first results were presented by Berger in Monte San Salvatore in Switzerland, with a median of 30kA[6] [2], comparable with those of countries in non-tropical zones. The first data of the magnitude of lightning return current obtained in tropical zone was recorded by Lee in Malaysia, which showed a median of 36kA[2].

In the tropical zone of South America, measurements were made in Brazil and Colombia, in the stations of Morro Cachimbo and Ilyapa, respectively. Professor Torres in Colombia obtained a median peak current of 42.9kA[2]. Table 1 shows the median current in different countries.

3. Ground flash density in Colombia and Brazil

According to IEC 62305-2, the Ground Flash Density (GFD) is the number of lightning flashes per km^2 per year. In many areas of the world, this value can be derived from data provided by Lighting Location Systems (LLS)[7]. The GFD was initially studied in Colombia by the Research Program on Acquisition and Analysis of Signals of the National University of Colombia (PAAS-UN)[4]. Currently, the company Keraunos S.A.S. is responsible for supplying the data corresponding to the GFD. The company use the network LINET, that has 10 magnetic field antennas located in different points of the country. This network has an efficiency over 90 percent[8]. The map of GFD in Colombia, published by the company Keraunos S.A.S. for the year 2014, is shown in figure 1.

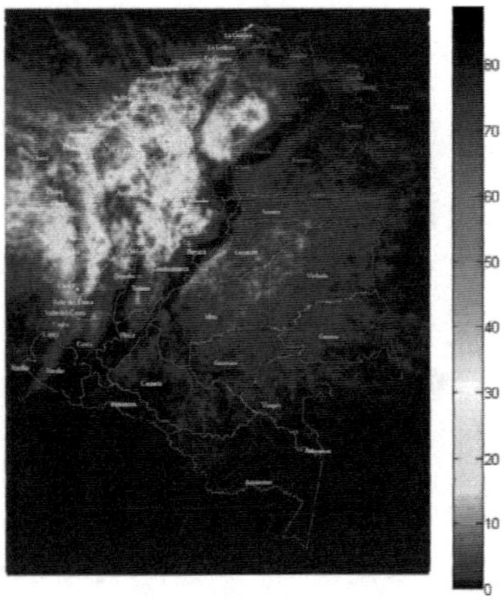

Figure 1. Ground Flash Density in Colombia 2014. [4]

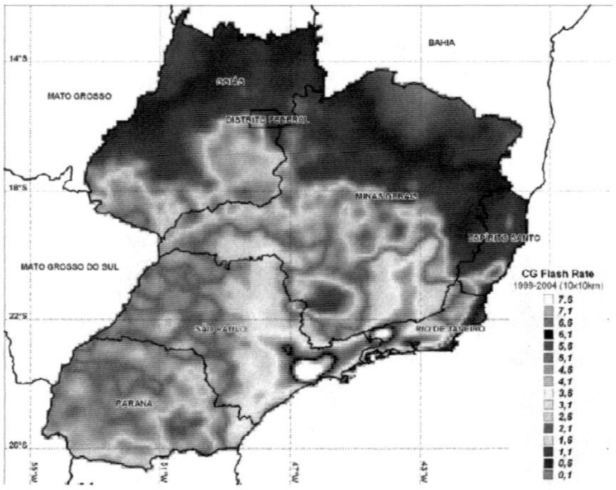

Figure 2. Ground Flash Density in Brazil. [11]

Colombia has areas with a GFD greater than 80 $\frac{strokes}{km^2\ year}$. This values are higher than in other countries like Brazil [10],the highest GFD in Brazil is about 8 $\frac{strokes}{km^2\ year}$ and it is in areas of Sao Paulo and Rio de Janeiro, data of the GFD in Brazil are obtained from the Integrated Lightning Detection Network (RINDAT), which is used to determine Brazilian standards for lightning protection [11], the map of GFD to Brazil is shown in Figure 2.

4. Types of lightning discharges

The literature of atmospheric electric discharges concludes that there are four types of lightning discharges, which are presented in next figure [12], where about 80 % of the total lightning are discharges that occur intracloud or cloud-to-cloud and only 20 % of the total discharges are cloud-to-ground discharges. [13].

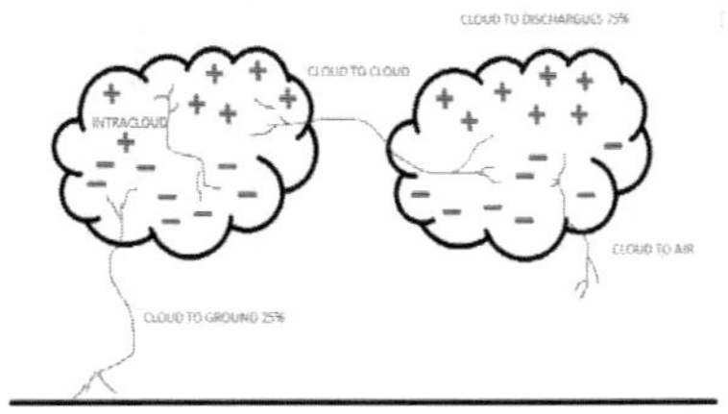

Figure 3. Types of lightning discharges.

With respect to cloud-to-ground discharges, we can say that there are different types of cloud-to-ground lightning discharges, which are downward negative lightning, upward negative lightning, downward positive lightning and upward positive lightning over cloud-to-ground discharges [14], the following figure shows each of the Types of cloud-to-ground Lightning

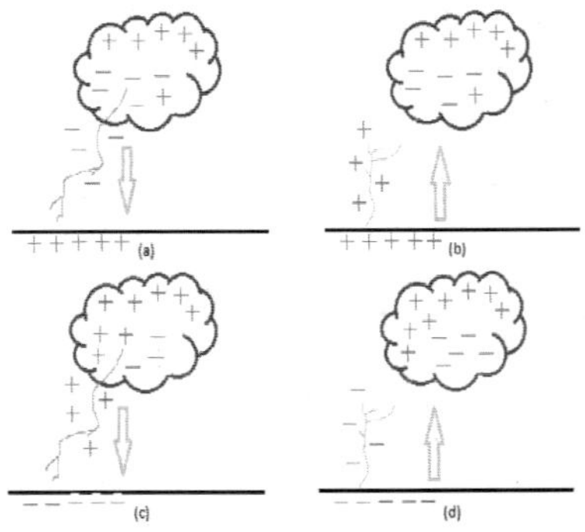

Figure 4. Types of Cloud to Ground Lightning. (a) Downward negative lightning, (b) Upward negative lightning, (c) Downward positive lightning, (d) Upward positive lightning

According to studies and statistics it is said that 90 % of the total Cloud-to-Ground Lightning Discharges are negative and only 10% positive [12]. Another thing that has been demonstrated with measurements and studies is that downward positive lightning current magnitudes are greater and destructive than those presented by downward negative lightning [2, 15]

5. Direct measurement of the lightning return current

A direct lightning measuring station must withstand the nominal values of an electric discharge, this considering the location of the station and the values that can be presented. The elements that must have a direct measurement station should be a derivative instrument, Rogowski coil, Shunt resistor, Pearson transformer and recording instruments as oscilloscopes [4].

In Colombia, a direct lightning metering station was implemented, which was called "experimental lightning direct measurement station, Ilyapa". This station was located in Samaná (Caldas), a point of high atmospheric electrical activity that was determined from the Information obtained from indirect measurements through Colombian LLS systems. This station was implemented with three objectives: to obtain direct measurements of lightning parameters, to compare the magnitudes of the lightning parameters with other latitudes and to improve lightning protection systems [4].

In Brazil, as in Colombia, a direct lightning metering station was implemented, this station was located in Morro do Cachimbo, this station was instrumented in 1985 at 15 km of Belo Horizonte [16], with this station has been able to measure more than 157 strokes in more than 13 years of operation [17].

6. Heidler's equation of lightning return current

Professor Heidler proposed the following equation that is currently used in IEC 62305:

$$i(t) = \frac{i_{max}}{k} * \frac{(t/\tau_1)^n}{1+(t/\tau_1)^n} * e^{(-t/\tau_2)} \tag{1}$$

Where, imax is the peak current, t is the time, k is the correction factor of the peak current, the coefficients y are the front time constant and the tail time constant respectively [7] [18].

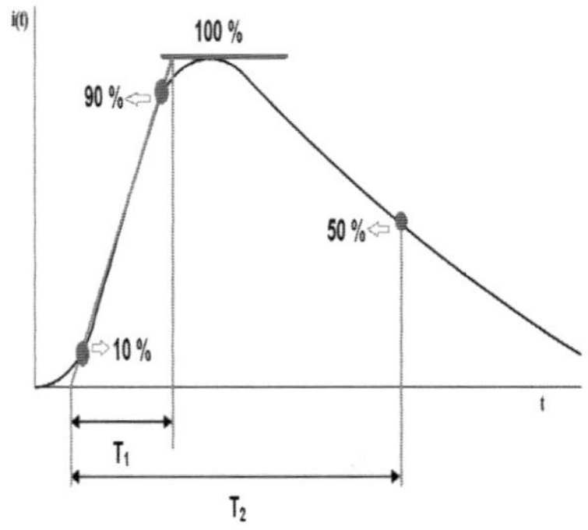

Figure 5. Definition of the short stroke current where T1 is the front time and T2 is the tail time.

For a short-stroke current, the signal has a front time T1 and a tail time T2, for the particular case of this article, the constants T1, T2, k and n will be found mathematically in order to adjust and minimize the error with respect to the direct lightning signals in Colombia and Brazil. Then, the behavior of Heiler equation will be observed with respect to measurements taken directly and compared with those obtained with the formula found in this article.

7. Direct measurement signals of the lightning return current for verifying the model

For the development of this article, three signs of verification were considered. Two of them were taken in Colombia and one in Brazil, with different parameters.

The parameters of each signal are totally different in front time, decay time and peak current, which makes them suitable to make the comparison of the formula of Heidler and the one presented in this article in different conditions.

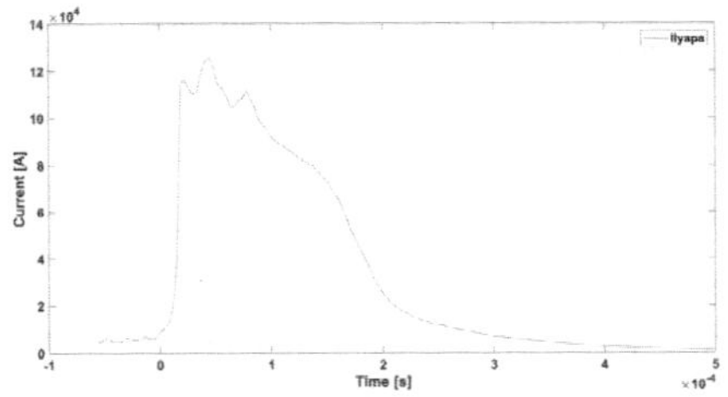

Figure 6. First lightning signal from Ilyapa, front time 0.00000074[s], decay time 0.0001622 [s] and peak current 125714 [A].

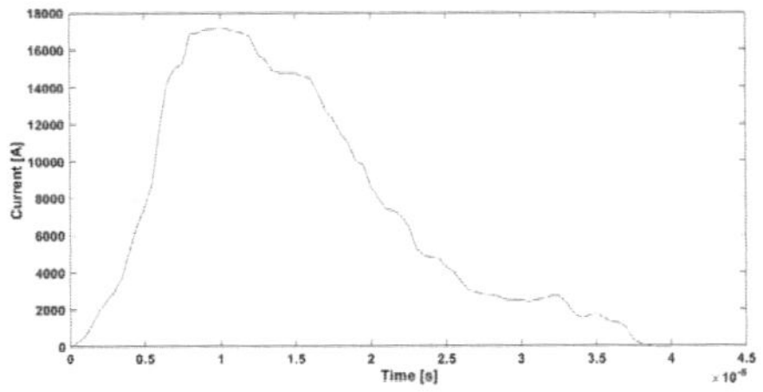

Figure 7. Second lightning signal from Ilyapa, front time 0.0000057 [s], decay time 0.000013 [s] and peak current 17190 [A].

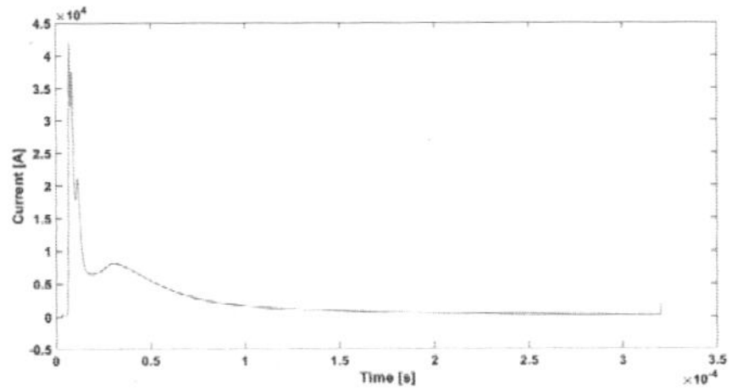

Figure 8. Lightning signal from Morro Cachimbo, front time 0.0000057 [s], decay time 0.0000189 [s] and peak current 42000 [A].

8. Mathematical modeling

The model is analyzed in three parts: signal rise, decay and amplitude.

8.1. Signal rise

The initial part of the typical graph of a lightning corresponds to the rise from t=0 to the peak current time. We use a pulse type function with stabilization to model it, as shown below:

$$\frac{\left(\frac{t}{t_u}\right)^{C_c}}{1+\left(\frac{t}{t_u}\right)^{C_c}} \tag{2}$$

The previous function is analyzed considering that t_u is the ascendant time, which is the time from t = 0 to the time it takes the signal to reach 90 % of the peak current as shown in the next figure.

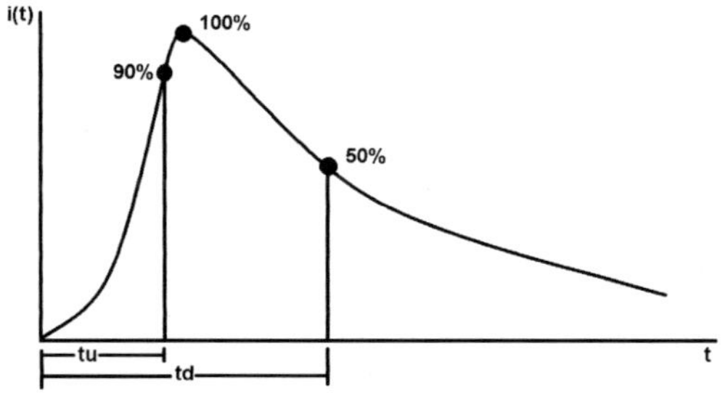

Figure 9. Ascendant time and decay time.

In equation 2, the time t_u of different signals is replaced and the coefficient Cc is changed with different values. Then, the results with a wave of 1.2 us of rising time and 50 us of decay time are shown with different coefficients Cc.

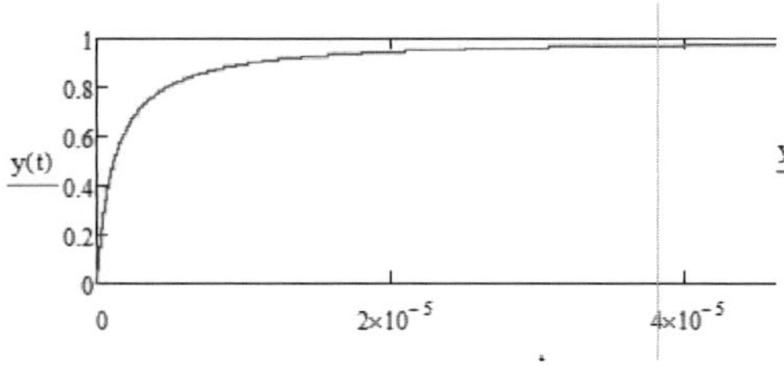

Figure 10. Ascendant time with Cc=1, y(t) is a equation 2.

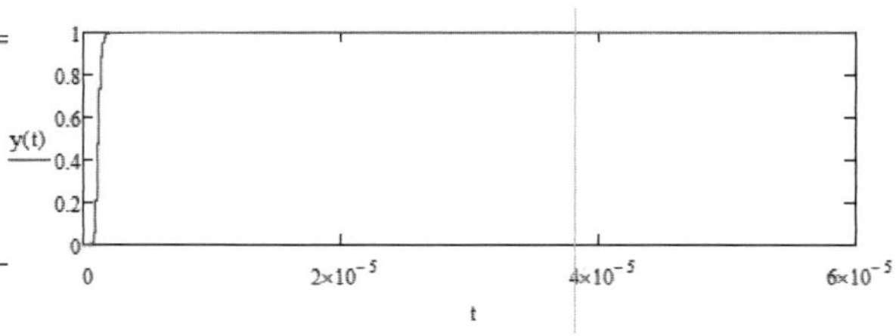

Figure 11. Ascendant time with Cc=10, y(t) is a equation 2.

From the above graphs, it can be determined that when Cc increases, the signal stabilization is faster. Now, the problem is that, when evaluating equation 2 with each value of tu, we did not find the value 0.9 of the magnitude. In the case of Cc=1 and Cc=10 in a time of 1.2 us, we found that in both cases f (1.2us)= 0.5, which is only a 50 % of the signal.

Therefore, we propose to modify equation 2 and find a new term, as shown in equation 3. Evaluating each value of Cc at time tu of the different signals, the error can be minimized and closer to 90 % .

$$\frac{\left(\frac{t+(t_u\ X)}{t_u}\right)^{C_c}}{1+\left(\frac{t+(t_u\ X)}{t_u}\right)^{C_c}}-0.9=0 \tag{3}$$

The solution of equation 3 was performed using Newton's numerical method and statistical methods, which determined that the proper value of X is 0.2338. This solution has an error of 0.0000001 % . Thus, the first part of the equation will be:

$$\frac{\left(\frac{t+(0.2338t_u)}{t_u}\right)^{C_c}}{1+\left(\frac{t+(0.2338\ \ t_u)}{t_u}\right)^{C_c}} \tag{4}$$

Equation 4 ensured that when Cc is greater than 10, the function reaches 90 % as predicted. Consequently, when the values of Cc are less than 10, the function does not reach 90 % , but the error is minimum compared to the results obtained in equation 2. For instance, even when Cc=1, the function reaches 57 % , that can be considered the worst case, and improves the 7 % of the equation 2. Nevertheless, this condition will be compensated during the later development of the model. Now we present the graphs obtained with equation 4:

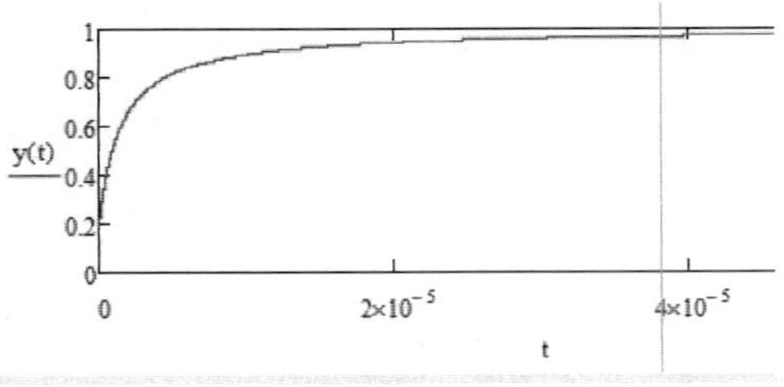

Figure 12. Ascendant time with Cc=1, y(t) is a equation 4.

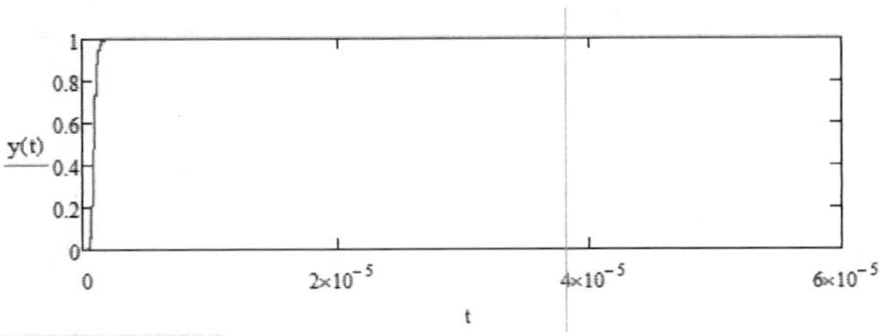

Figure 13. Ascendant time with Cc=10, y(t) is a equation 4.

The next step for the development of the model is to transfer the function to zero. Thus, we evaluate the function at zero for each value of Cc to have a reference point:

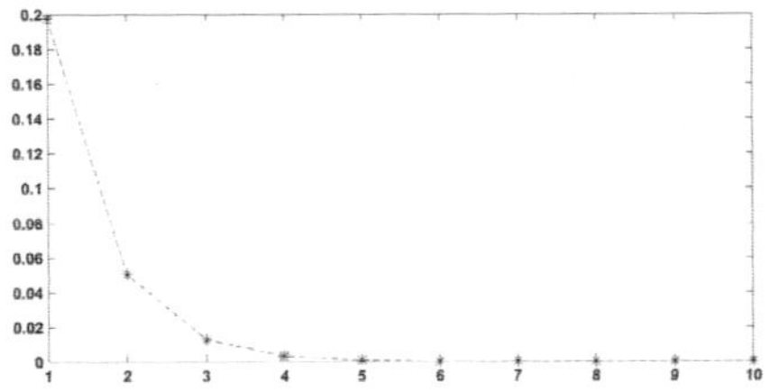

Figure 14. Vertical offset based on Cc, y(t) is a equation 4 * X, where X is a new term for transfer this function to zero. Equation 5, have two terms, the left part is a equation 4 and right equation is a term determined in this graph like offset.

The previous Figure, $0.7773\ e^{-}1.37Cc$ was determined as the needed term to transfer the signal to zero. The term slightly affects the signal. 90 % is kept when it is evaluated in tu, with Cc=10. In contrast, the signal is decreased to 51 % when Cc=1. The function with the term previously found is presented below:

$$\left(\frac{\left(\frac{t+0.2338*t_u}{t_u}\right)^{C_c}}{1+\left(\frac{t+(0.2338*t_u)}{t_u}\right)^{C_n}} - 0.7773*e^{-1.37*\ C_c}\right) \tag{5}$$

Journal of Physics: Conference Series **2135** (2021) 012007 doi:10.1088/1742-6596/2135/1/012007

The behavior of the above function is shown below with values of Cc=1 and 10, and an ascendant time of 1.2us.

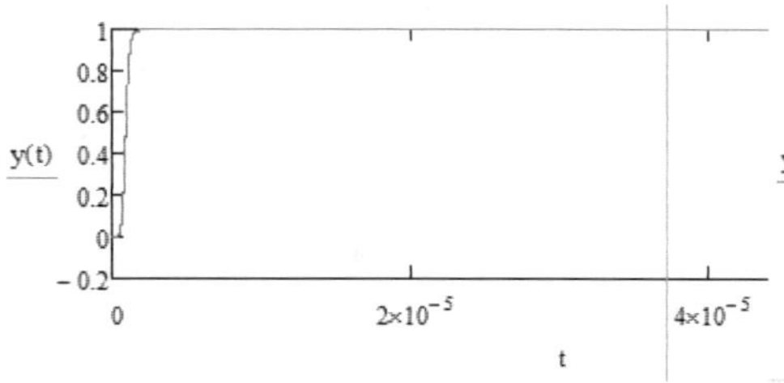

Figure 15. Front signal in equation 5 with Cc=10.

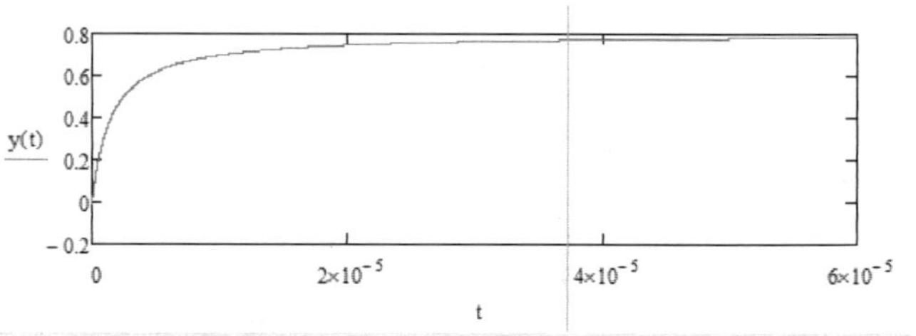

Figure 16. Front signal in equation 5 with Cc=1.

The equation found for the rise of the signal works properly with values greater than Cc=4. With the different values calculated in tu, the signal is greater than 80 % . We will try to increase this value with the next procedures.

8.2. Decrease in signal

Once the approach of the front of the model has been completed, the decay of the model is proceeded. The decreasing exponential is used as shown in equation 6:

$$e^{\frac{-t}{t_d}} \tag{6}$$

Equation 6 presents an ideal behavior of a decrease. However, the function must be transferred on the horizontal axis, so that it starts in the value of the ascendant time, and ensure the 90 % obtained in the rise of the curve. The graph of the equation 7 was done using the decay time and the ascendant time with 1.2us and 50 us respectively. The modification of the equation 6 is shown in the following function:

$$e^{\frac{-(t-t_s)}{t_d}} \tag{7}$$

The graph of next figure shows that the equivalent percentage of the signal in the decay time td is 39 % .

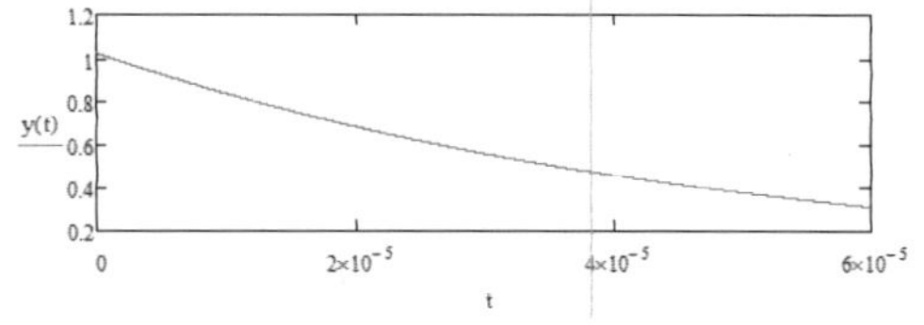

Figure 17. Signal decay, y(t) is a equation 7.

The possibility of transferring the model to the ascendant time in the equation 5 and compacting it with the equation 7 ensures that a front signal between 62 % and 90 % in the ascendant time. 62 % was obtained with Cc=1 and 90 % with Cc=10. The compact equation of ascendant and decay times (8) and its correspondent next 2 graphs are shown below.

$$e^{\frac{-(t-t_u)}{t_d}} * \left(\frac{\left(\frac{t+0.2338*t_u}{t_u}\right)^{Cc}}{1+\left(\frac{t+(0.2338*t_u)}{t_u}\right)^{Cc}} - 0.7773 * e^{-1.37*C_c} \right) \tag{8}$$

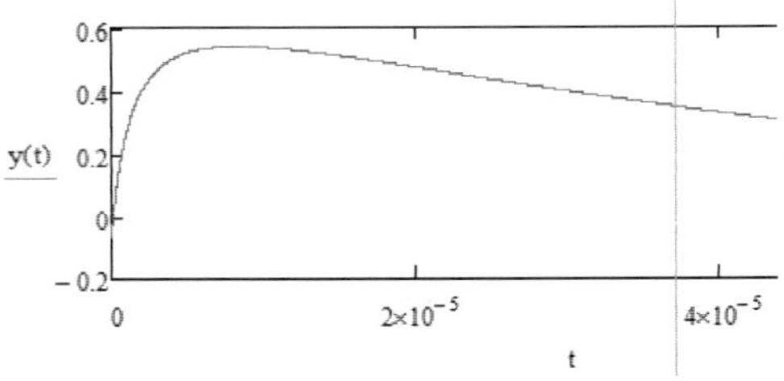

Figure 18. Graph of the function 8, with Cc=10.Percentage of the signal at time tu = 90 % , Percentage of the signal at time td =38.5 %

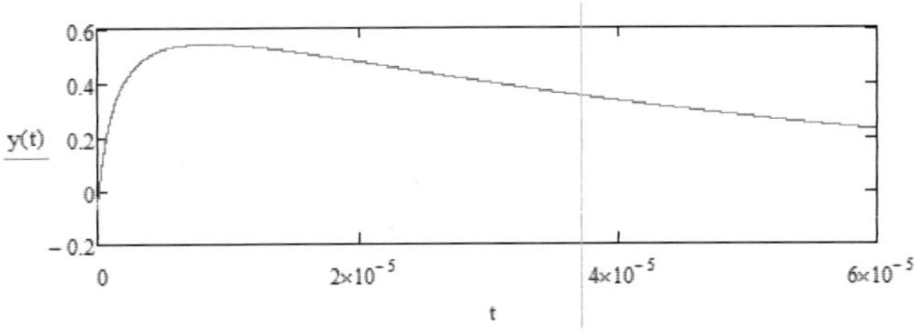

Figure 19. Graph of the function 8, with Cc=1.Percentage of the signal at time tu = 61 % , Percentage of the signal at time td =50 %

The function found to model the decay time behaves properly between a range of 38.5 % and 50 % , where the approach value increases when Cc is lower.

8.3. Relationship between ascendant time and decay time, Cc

So far, the rise of the signal and its decay have been modeled independently. Now we are going to propose a relation between both through the term Cc, which is going to be called coefficient of curvature. This coefficient Cc relates the rise of the signal with the decay by means of a new term called curvature time tc, which is the time it takes the 50 % of the rise wave to reach the 50 % of the decay signal, as shown in the next figure.

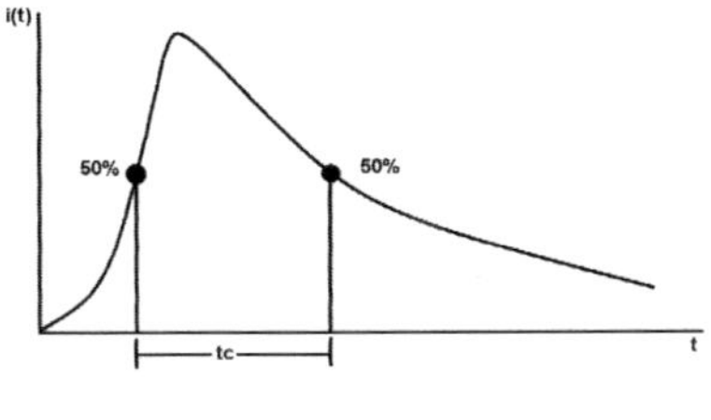

Figure 20. curvature time Cc.

When the term Cc was found to relate the rise and the decay signal as represented in equation 9, a correction had to be made at the decay signal term.

$$C_c = \left(\frac{2t_c}{1 * 10^{-5}} \right) \tag{9}$$

Considering the coefficient of curvature, the adaptation of the decay signal was made, as shown in the following figure.

The coefficient found for the correction is 0.1952 ln(Cc) + 0.3274. It was coupled in equation 10 as follows:

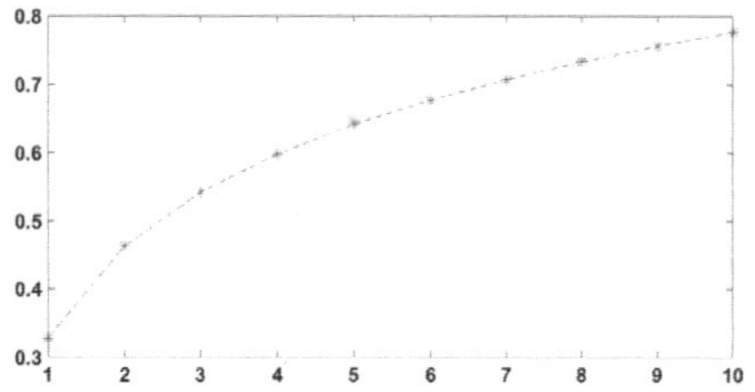

Figure 21. Decay signal correction, Y axis is Cc and X axis is tc.

$$e^{\frac{-(t-t_u)}{t_d(0.3274+0.1952*\ln(C_c))}} * \left(\frac{\left(\frac{t+0.2338*t_u}{t_u}\right)^{C_c}}{1+\left(\frac{t+(0.2338*t_u)}{t_u}\right)^{C_c}} - 0.7773*e^{-1.37*\ C_c} \right) \quad (10)$$

In the implementation of the equation, the term Cc can be used with real signals as shown in equation 9. However, the values of Cc= 10 and Cc = 1 are assumed for testing previous equation.

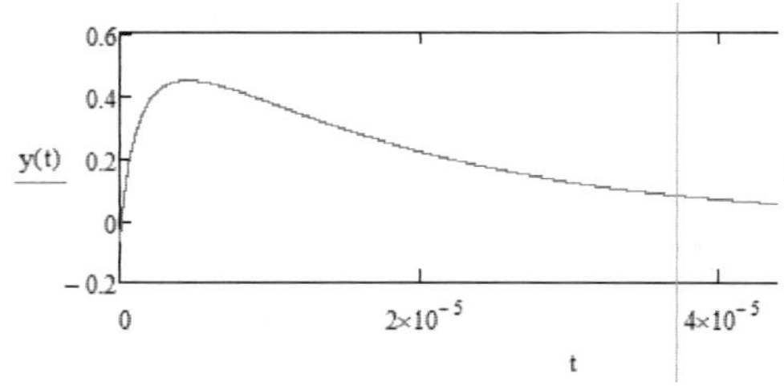

Figure 22. Graph of equation 10 with Cc=1.Percentage of the signal at time tu= 71.5 % , Percentage of the signal at time td= 9.5 %

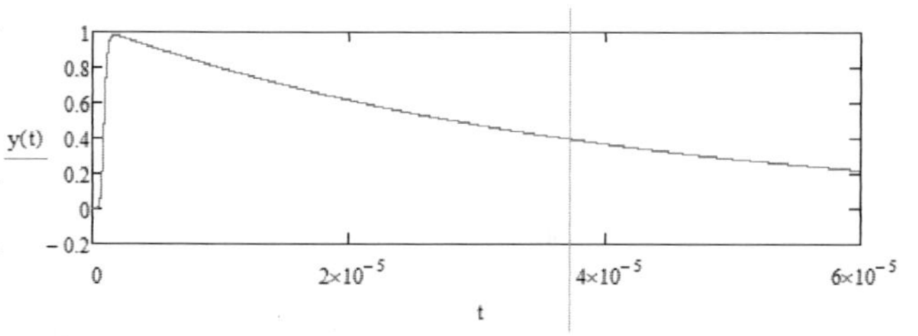

Figure 23. Graph of equation 10 with Cc=10.Percentage of the signal at time tu= 90 % , Percentage of the signal at time td= 36 %

Now, equation10 is solved and its results are compared with the results obtained in equation 8 with a signal of 1.2 / 50 us and Cc = 1.

Results went from 61 % in equation 8, to 71.5 % in equation 10, evaluating the ascendant time. The signal had an improvement of 10.5 % . However, when the decay time was evaluated, it went from 50 % in equation 8, to 9.5 % in equation 10, which means a loss of 40.5 % . Alternatively, when the signal is evaluated with Cc= 10, the ascendant time kept 90 % in both equations, but the decay time went from 38.5 % in equation 8, to 36 % in equation 10, which is a loss of 2.5% .

According to this, it is possible to determine that Equation 10 improved the percentages of the ascendant time in cases where Cc was small, and maintain them when Cc had large values. On the contrary, the decay time was lower in all cases. Nevertheless, the signal front is more important in lightning signals. Therefore, this aspect was improved and an expression for the coefficient Cc could be found.

8.4. Signal amplitude

For the signal amplitude, the maximum peak of a signal to be analyzed is used, but an additional coefficient must be applied to adapt the signal as a function of Cc. The graph obtained from the adaptation data as a function of Cc is shown below:

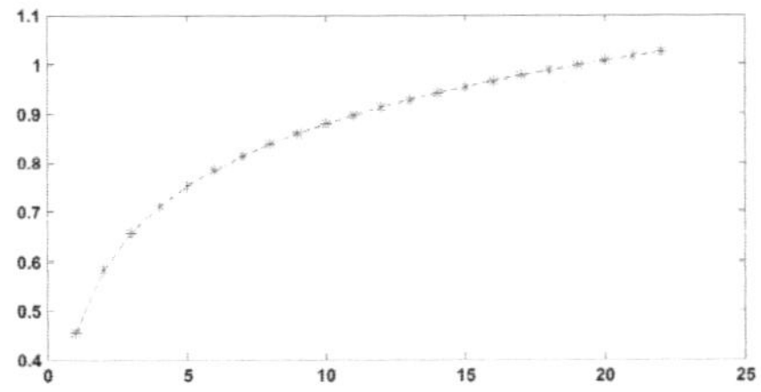

Figure 24. Amplitude factor.

$$i(t) = \left(\frac{A}{0.4545 + 0.1847\ln\left(C_c\right)}\right) * \left(e^{\frac{-(t-t_u)}{t_d(0.3274+0.1952\ln(C_c))}}\right) * \left(\frac{\left(\frac{t+0.2338*t_u}{t_u}\right)^{C_c}}{1+\left(\frac{t+(0.2338*t_u)}{t_u}\right)^{C_c}} - 0.7773 * e^{-1.37* \ C_c}\right) \quad (11)$$

Remembering that, Cc $= \frac{2tc}{1x10^{-5}}$

9. Mathematical model verification

The verification of the model found is maded, contrasting it with the Heidler equation, for which both equations are applied to the three signals presented in section 7. Making the contrast of Heidler's equation with that found in this model will determine the advantages and disadvantages of the model found.

9.1. Verification with Lightning Ilyapa signal

The Ilyapa signal parameters for the model found are shown below:

$$C_c = \frac{2 * t_c}{1x10^{-5}}$$

$$t_c = 145,0x10^{-6}s$$

$$C_c = \frac{2 * 145,0x10^{-6}}{1x10^{-5}} = 29$$

$$C_c = 29$$

$$t_u = 4x10^{-5}[s]$$

$$t_d = 1,626x10^{-4}[s]$$

With the defined parameters, the following three signals are plotted:

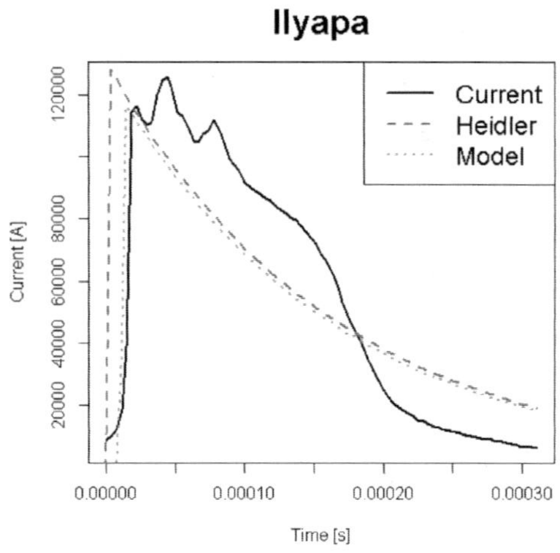

Figure 25. Graph of the Ilyapa signal, comparison, between Heidler and model found.

9.2. Verification with Lightning Ilyapa signal 2

The Ilyapa signal 2 parameters for the model found are shown below:

$$C_c = \frac{2 * t_c}{1x10^{-5}}$$

$$t_c = 2,09x10^{-6}s$$

$$C_c = \frac{2 * 2,09x10^{-6}}{1x10^{-5}}$$

$$C_c = 4,18$$

$$t_u = 9x10^{-6}[s]$$

$$t_d = 2,0004x10^{-5}[s]$$

With the defined parameters, the following three signals are plotted:

9.3. Verification with Lightning Morro Cachimbo signal

The Morro Cachimbo signal parameters for the model found are shown below:

$$C_c = \frac{2 * t_c}{1x10^{-5}}$$

$$t_c = 70,02x10^{-6}s$$

$$C_c = \frac{2 * 70,02x10^{-6}}{1x10^{-5}}$$

$$C_c = 14$$

$$t_u = 6,474x10^{-6}[s]$$

$$t_d = 9,6102x10^{-6}[s]$$

With the defined parameters, the following three signals are plotted:

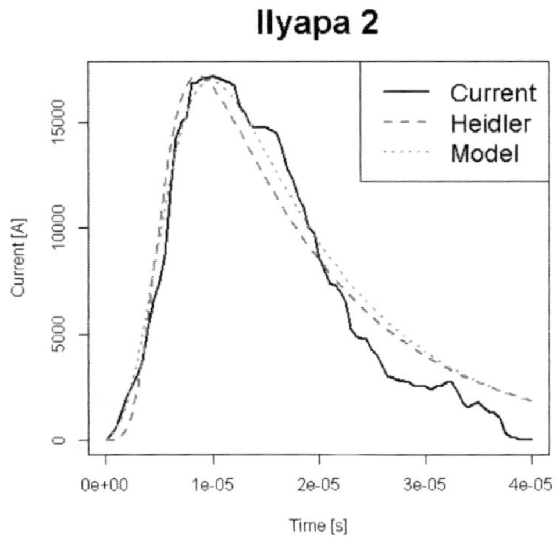

Figure 26. Graph of the Ilyapa signal 2, comparison, between Heidler and model found.

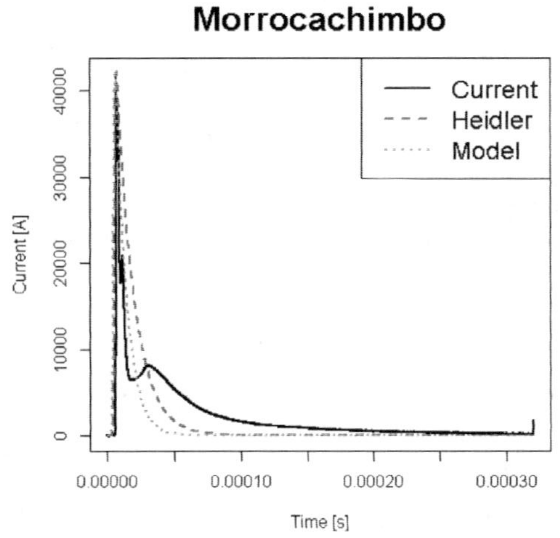

Figure 27. Graph of the Morro Cachimbo, comparison, between Heidler and model found.

10. Correlation

Here we use the Spearman's correlation coefficient, since it is non-parametric and allows measuring relation between non-normal samples, in this case the Pearson's correlation coefficient is not applicable because since data does not accomplish assumptions. Bellow we show the Spearman's correlation for each one of the signals:

Table 1. Correlation Ilyapa signal 1.

Method	Value
Heidler	0.8580
Model	0.9677

Table 2. Correlation Ilyapa signal 2.

Method	Value
Heidler	0.9730
Model	0.9878

Table 3. Correlation Morro Cachimbo signal.

Method	Value
Heidler	0.9500
Model	0.9174

The correlation for the first signal, the equation presented in this paper obtains a better correlation respect to real signal, in the second correlation for the Ilyapa signal 2, the equation presented in this paper obtains a better correlation respect to real signal, respecto to correlation of Morro do Cachimbo the signal of the Heidler model presents a proper behavior as particularly can be seen in the second subsequent discharge, which if had not been present the model obtained in this paper would have had an ideal performance.

11. Conclusion

With respect to the first approximation of the mathematical model found for the return current applied to real signals, it can be said that it fits well to the curves of the different signals, especially in the front of the signal.

The coefficient C correlates well in the model found and can be used in the Heidler equation, this value specifically might to be helpful for replicate a real signal.

The curvature time t is an easy parameter to find and important to recreate the signal.

The variables of the equation are A, tu, td and tc, all these variables are dates of a direct signal.

The importance of applying the Cc coefficient either in the model found in this paper or in Heidler's equation is to be able to simulate real signals.

Each variable necessary for the equation finding in this paper are easily calculates for replicate real signals.

References

[1] J. Galvin and C. Jones, *The weather and climate of the tropics: Part 10 – Tropical agriculture*, Weather, vol 64, pp. 156-161. , June, 2009.

[2] H. Torres, *Protección contra rayos*, National University of Colombia and Icontec, second edition, pp. 35-50. , Bogotá, October, 2010.

[3] J. Rubiano, *Análisis estadístico de dos parámetros de rayo en zona tropical*, Francisco José de Caldas District University, Thesis, pp. 26-27. , Bogotá, June, 2016.

[4] H. Torres, *El Rayo. Mitos, leyendas, ciencia y tecnología*, National University of Colombia, Second edition, Chapter 4, pp. 111-171. , Bogotá, 2002.
[5] F. Heidler, *Parameters of lightning current given in IEC 62305 - background, experience and outlook*, 29th International Conference on lightning protection, pp. 1-22. , Uppsala, Sweden, June, 2008.
[6] M. Uman, *The Lightning Discharge*, Academic Press INC., vol 39, pp. 32-54. , Orlando San Diego, EE.UU., 1987.
[7] IEC, *IEC 62305, Protection against lightning – Risk management*, International Electrotechnical Commission, part 2, Ed 3,. , Geneva, Switzerland, 2005.
[8] M. Uman, *K Services*, Keraunos science of lightning, Keraunos S.A.S., http://keraunos.co/index.php/es-co/. , Bogotá, Colombia, March, 2017.
[9] NASA, *National Aeronautics and Space Administration*, https://ghrc.nsstc.nasa.gov/lightning/images/browse/mission.png, Department of Meteorology. , 2015.
[10] G. Huffines and R. Orville, *Lightning Ground Flash Density and Thunderstorm Duration in the Continental United States: 1989-96*, Cooperative Institute for Applied Meteorological Studies, Department of Meteorology. Texas, pp. 7. , 1998.
[11] I. Pinto and O. Pinto and K. Naccarato, *HOW GROUND FLASH DENSITY OBTAINED BY LIGHTNING LOCATION NETWORKS CAN BE USED IN LIGHTNING PROTECTION STANDARDS: A CASE STUDY IN BRAZIL*, 19th International Lightning Detection Conference, pp. 1-3. , 2006.
[12] V. Rakov, *Fundamentals of Lightning*, International Symposium on Lightning Protection, pp. 4-7. , Kathmandu, Nepal, October, 2011.
[13] B.Kucieńska , *Los misterios de la nube de tormenta*, Seminario de divulgacipon del centro de ciencias de la atmósfera, Universidad Nacional Autónoma de México, pp. 18-21. , City of Mexico, Mexico,13, October,2013.
[14] V. Rakov, *Lightning Parameters od engineering interest: Application of lightning detection technologies*, International Symposium on Lightning Protection, pp. 10-13. , Bangkok, Thailand,7, November, 2012.
[15] O. Pinto and L. Pinto, *Tormentas Positivas : Sorpresa en los Cielos Brasileños*, Revista de Divulgación Científica y Tecnológica de la Asociación Ciencia Hoy, Vol 8,No 44. , Argentina, January, 1998.
[16] M. Rubinstein and M. Paolone and C. Romero and F. Rachidi, *Instrumentation of the Säntis Tower for Lightning Current Measuremen*, ÉCOLE POLYTECHNIQUE FÉDÉRALE DE LAUSANNE. , January, 2012.
[17] S. Visacro and A. Soares, *Statistical analysis of lightning current parameters: Measurements at Morro do Cachimbo Station*, JOURNAL OF GEOPHYSICAL RESEARCH, vol 109,pp. 5-12.. , Argentina,10, January, 2004.
[18] F. Heidler and Z. Filisowski, *Parameters of lightning current given in IEC 62305 - background, experience and outlook*, 29th International Conference on lightning protection, pp. 4-20. , Upssala, Sweden,23-26, June, 2008.

Computational Tool for Determining the Model in Sequence Components of Overhead Transmission Lines

Luis Imbachi Guerrero, Fredy Jiménez Rubio, Mario Rodríguez Barrera, Diego Giral Ramírez
Universidad Distrital Francisco José de Caldas. Colombia.
lfimbachig@correo.udistrital.edu.co, fajimenezr@correo.udistrital.edu.co, marodriguezb@udistrital.edu.co, dagiralr@udistrital.edu.co

Abstract. An indispensable element in addressing the current problem of non-ionizing electromagnetic pollution in the environment is a review of the levels of exposure to the electric and magnetic fields produced by the lines of electric power transmission and distribution systems. In order to establish the exposure levels, it is necessary to determine the model of the lines. Considering that a computational simulation is a helpful tool for power system analysis, this article presents a computational tool developed in Matlab App Designer for the model-in-sequence components of the parameters that make up a transmission line. This tool allows the user to work in a friendly and parameterizable environment according to the performed tests. In order to verify the tool's performance, two case studies are implemented. The first one is for a transposed transmission line and the second one for a non-transposed transmission line. The results obtained are compared with commercial software, acquiring a maximum error of 0.16402 %.

1. Introduction

In order to supply electric power to a given territory, it is necessary to implement an electric power system, which is made up of a generation, transmission, and distribution stage [1]–[3]. A transmission system is mainly formed by overhead lines that transport power to a large part of the territories [4]–[7]. Thus, it is of utmost importance to analyze the electrical parameters and characteristics in the conductors of the lines since these include impedance and admittance parameters, which are essential for load flow analysis, short-circuit studies, coordination of protections, among others [8]–[10].

Based on the above, it is considered essential to determine the model in sequence components of the parameters that make up a transmission line [8], [9]. This article aims to present a computational tool developed in Matlab App Designer for transmission line modeling. The results are compared with the software for the study of power systems NEPLAN.

As there is significant interest in the modeling of transmission lines, computational tools have been developed to facilitate the learning process or accompany it as a working tool [11], [12]. The following are the most relevant documents associated with the development of non-commercial simulation tools.

In [13], the authors present a computer program called CELTIO developed with MATLAB software, which allows simulating the electric field of a transmission line under the finite element method, which is applied to electrostatic problems that allow obtaining the electric field induced under a transmission line.

In [6], the authors relate the methods for the calculation of basic electrical parameters in a transmission line, which are resistance, inductance, and capacitance. The authors of the document developed a computational tool in MATLAB with a graphical interface named CPEL, which is responsible for guiding the user in the process of determining the values of the electrical parameters of a transmission line. Likewise, the authors reflect in the document explanatory exercises to give an accompaniment to the reader.

This paper is organized and presented in five sections. Section 2 describes the mathematical formulation. Section 3 presents the methodology describing the modules that make up the application. Section 4 analyzes the results obtained for two case studies. Section 5 establishes the general conclusions of the work.

2. Mathematical Formulation

The software developed is focused on the imaging method. This method allows determining the series model and the capacitances per phase; later, by applying the Kron reduction and the similarity transformation, the equivalent in sequence components, and finally, the equivalent circuit is obtained. The following is the mathematical formulation for the modeling of non-transposed transmission lines, a formulation that is easily adaptable to a transposed transmission line.[14]–[16].

2.1. Non-transposed line: Matrix Z_{abc} (Series model per phase)

In a non-transposed transmission line with n guard wires, the self-impedances $\mathbf{Z_{ii}}$ are equal; the impedances $\mathbf{Z_{ij}}$ are different from each other, their calculation is done by equations (2) and (3), for the remaining positions of the matrix $\mathbf{Z_{abc}}$ use is made of equations (4), (5), (6), (6), (7), and (8) as appropriate.

$$[Z_{abc}] = \begin{bmatrix} Z_{aa} & Z_{ab} & Z_{ac} & Z_{au} & \cdots \\ Z_{ba} & Z_{bb} & Z_{bc} & Z_{bu} & \cdots \\ Z_{ca} & Z_{cb} & Z_{cc} & Z_{cu} & \cdots \\ Z_{ua} & Z_{ub} & Z_{uc} & Z_{uu} & \cdots \\ \vdots & \vdots & \vdots & \vdots & \ddots \end{bmatrix} \tag{1}$$

$$Z_{ii} = \left[(r_a + r_e) + j0{,}0754 \cdot \ln\left(\frac{D_e}{RMG}\right)\right]\left[\frac{\Omega}{km}\right] \tag{2}$$

$$Z_{ij} = \left[r_e + j0{,}0754 \cdot \ln\left(\frac{D_e}{D_{ij}}\right)\right]\left[\frac{\Omega}{km}\right] \tag{3}$$

$$Z_{au} = Z_{ua} = \left[r_e + j0{,}0754 \cdot \ln\left(\frac{D_e}{D_{au}}\right)\right]\left[\frac{\Omega}{km}\right] \tag{4}$$

$$Z_{bu} = Z_{ub} = \left[r_e + j0{,}0754 \cdot \ln\left(\frac{D_e}{D_{bu}}\right)\right]\left[\frac{\Omega}{km}\right] \tag{5}$$

$$Z_{cu} = Z_{uc} = \left[r_e + j0{,}0754 \cdot \ln\left(\frac{D_e}{D_{cu}}\right)\right]\left[\frac{\Omega}{km}\right] \tag{6}$$

$$Z_{uu} = \left[(r_u + r_e) + j0{,}0754 \cdot \ln\left(\frac{D_e}{RMG_u}\right)\right]\left[\frac{\Omega}{km}\right] \tag{7}$$

$$Z_{su_{(3+m)n}} = \left[r_e + j0{,}0754 \cdot \ln\left(\frac{D_e}{D_{(3+m)n}}\right)\right]\left[\frac{\Omega}{km}\right] \tag{8}$$

Where D_{ij}: Distance between phase i and phase j

2.1.1. Kron Reduction Matrix Z_{abc}

$$\left[Z_{abc_{RK}}\right] = [Z_1] - [Z_2][Z_4]^{-1}[Z_3] \tag{9}$$

Where:

$$[Z_1]=\begin{bmatrix} Z_s & Z_m & Z_m \\ Z_m & Z_s & Z_m \\ Z_m & Z_m & Z_s \end{bmatrix} \quad [Z_2]=\begin{bmatrix} Z_{su} & \cdots \\ Z_{su} & \cdots \\ Z_{su} & \cdots \end{bmatrix} \quad [Z_3]=\begin{bmatrix} Z_{su} & Z_{su} & Z_{su} \\ \vdots & \vdots & \vdots \end{bmatrix} \quad [Z_4]=\begin{bmatrix} Z_{uu} & \cdots \\ \vdots & \ddots \end{bmatrix} \tag{10}$$

2.1.2. Matrix Z_{012} (Serial model in sequence components)

$$[Z_{012}]=[A]^{-1}\cdot\left[Z_{abc_{RK}}\right]\cdot[A]$$

$$[Z_{012}]=\begin{bmatrix} Z_s+2Z_m & 0 & 0 \\ 0 & Z_s-Z_m & 0 \\ 0 & 0 & Z_s-Z_m \end{bmatrix}\left[\frac{\Omega}{km}\right] \tag{11}$$

2.2. Matrix C_{abc} (Matrix of capacitances per phase)

In a transposed transmission line with n guard wires, the maxwell coefficients P_{ii} are equal; and the maxwell coefficients P_{ij} are different from each other, their calculation is performed through Equations (13) and (14), for the remaining positions of the matrix P_{abc} we make use of equations (15), (16), (17), (18) as appropriate.

$$[P_{abc}]=\begin{bmatrix} P_{aa} & P_{ab} & P_{ac} & P_{au} & \cdots \\ P_{ba} & P_{bb} & P_{bc} & P_{bu} & \cdots \\ P_{ca} & P_{cb} & P_{cc} & P_{cu} & \cdots \\ P_{ua} & P_{ub} & P_{uc} & P_{uu} & \cdots \\ \vdots & \vdots & \vdots & \vdots & \ddots \end{bmatrix} \tag{12}$$

$$P_{ii}=\frac{1}{2\pi\varepsilon_0}\ln\left(\frac{h_{ii}}{RMG_i}\right)\left[\frac{m}{F}\right] \tag{13}$$

$$P_{ij}=\frac{1}{2\pi\varepsilon_0}\ln\left(\frac{l_{ij}}{D_{ij}}\right)\left[\frac{m}{F}\right] \tag{14}$$

$$P_{uu}=\frac{1}{2\pi\varepsilon_0}\ln\left(\frac{h_u}{RMG_u}\right)\left[\frac{m}{F}\right] \tag{15}$$

$$P_{ua}=P_{au}=\frac{1}{2\pi\varepsilon_0}\ln\left(\frac{l_{ua}}{D_{ua}}\right)\left[\frac{m}{F}\right] \tag{16}$$

$$P_{ub}=P_{bu}=\frac{1}{2\pi\varepsilon_0}\ln\left(\frac{l_{ub}}{D_{ub}}\right)\left[\frac{m}{F}\right] \tag{17}$$

$$P_{uc}=P_{cu}=\frac{1}{2\pi\varepsilon_0}\ln\left(\frac{l_{uc}}{D_{uc}}\right)\left[\frac{m}{F}\right] \tag{18}$$

Where:
l_{iu}: Distance between the phases and the guard wire image.
h_u: Distance between guard wires and their image.
h: Distance between the phase and its image.
l: Distance between the phase and the image of the other phases.
h_{u_k}: Distance between the guard wire and its image.
D_{i_u}: Distance between the phase and the guard wire.
l_{i_u}: Distance between the phase and the image of the guard wire.

2.2.1. Kron Reduction Matrix P_{abc}

$$[P_{abc_{RK}}] = [P_1] - [P_2][P_4]^{-1}[P_3] \tag{19}$$

Where:

$$[P_1] = \begin{bmatrix} P_s & P_m & P_m \\ P_m & P_s & P_m \\ P_m & P_m & P_s \end{bmatrix} \quad [P_2] = \begin{bmatrix} P_{mu} & \cdots \\ P_{mu} & \cdots \\ P_{mu} & \cdots \end{bmatrix} \quad [P_3] = \begin{bmatrix} P_{um} & P_{um} & P_{um} \\ \vdots & \vdots & \vdots \end{bmatrix} \quad [P_4] = \begin{bmatrix} P_{su} & \cdots \\ \vdots & \ddots \end{bmatrix} \tag{20}$$

2.2.2. Matrix C_{abc} (Matrix of capacitances per phase)

$$[C_{abc}] = [P_{abc_{RK}}]^{-1} \tag{21}$$

2.2.3. Matrix C_{012} (Matrix of capacitances in sequence components)

$$[C_{012}] = [A]^{-1} \cdot [C_{abc}] \cdot [A]$$

$$[C_{012}] = \begin{bmatrix} C_0 & 0 & 0 \\ 0 & C_1 & 0 \\ 0 & 0 & C_2 \end{bmatrix} \left[\frac{\Omega}{km}\right] \tag{22}$$

Remark: In order to determine the model of sequence components in transposed lines, there are two alternatives. The first uses the same formulas, as for a non-transposed line, with the difference that for a transposed line the D_{ii}, h_{ii} and l_{ij} are changed by equivalent average distances. And the second way is to determine the average of the diagonal and off-diagonal elements as shown in equations (23) and (24).

$$[Z_{abc}]_{Transposed} = \begin{bmatrix} \frac{Z_{aa}+Z_{bb}+Z_{cc}}{3} & \frac{Z_{ab}+Z_{ca}+Z_{bc}}{3} & \frac{Z_{ab}+Z_{ca}+Z_{bc}}{3} \\ \frac{Z_{ab}+Z_{ca}+Z_{bc}}{3} & \frac{Z_{aa}+Z_{bb}+Z_{cc}}{3} & \frac{Z_{ab}+Z_{ca}+Z_{bc}}{3} \\ \frac{Z_{ab}+Z_{ca}+Z_{bc}}{3} & \frac{Z_{ab}+Z_{ca}+Z_{bc}}{3} & \frac{Z_{aa}+Z_{bb}+Z_{cc}}{3} \end{bmatrix} \tag{23}$$

$$[P_{abc}]_{Transposed} = \begin{bmatrix} \frac{P_{aa}+P_{bb}+P_{cc}}{3} & \frac{P_{ab}+P_{ca}+P_{bc}}{3} & \frac{P_{ab}+P_{ca}+P_{bc}}{3} \\ \frac{P_{ab}+P_{ca}+P_{bc}}{3} & \frac{P_{aa}+P_{bb}+P_{cc}}{3} & \frac{P_{ab}+P_{ca}+P_{bc}}{3} \\ \frac{P_{ab}+P_{ca}+P_{bc}}{3} & \frac{P_{ab}+P_{ca}+P_{bc}}{3} & \frac{P_{aa}+P_{bb}+P_{cc}}{3} \end{bmatrix}. \tag{24}$$

Remark: To determine $Z_{abc_{Rk}}$, Z_{012}, C_{abc} y C_{012} the same procedure is followed for a non-transposed line.

3. Methodology

The methodology used for the development of the tool is presented below. The first part describes the algorithms used for the operation process, and the second part describes the module structure used.

3.1. Operation process

The computational tool developed in Matlab App Designer allows the user to progressively advance in the configuration of the problem to be solved. This process is described below, using the flowchart shown in Figures 1, 2, and 3.

Figure 1 shows five operation blocks that represent the panels that the user must configure in the computational tool according to the information contained in the problem to be solved.

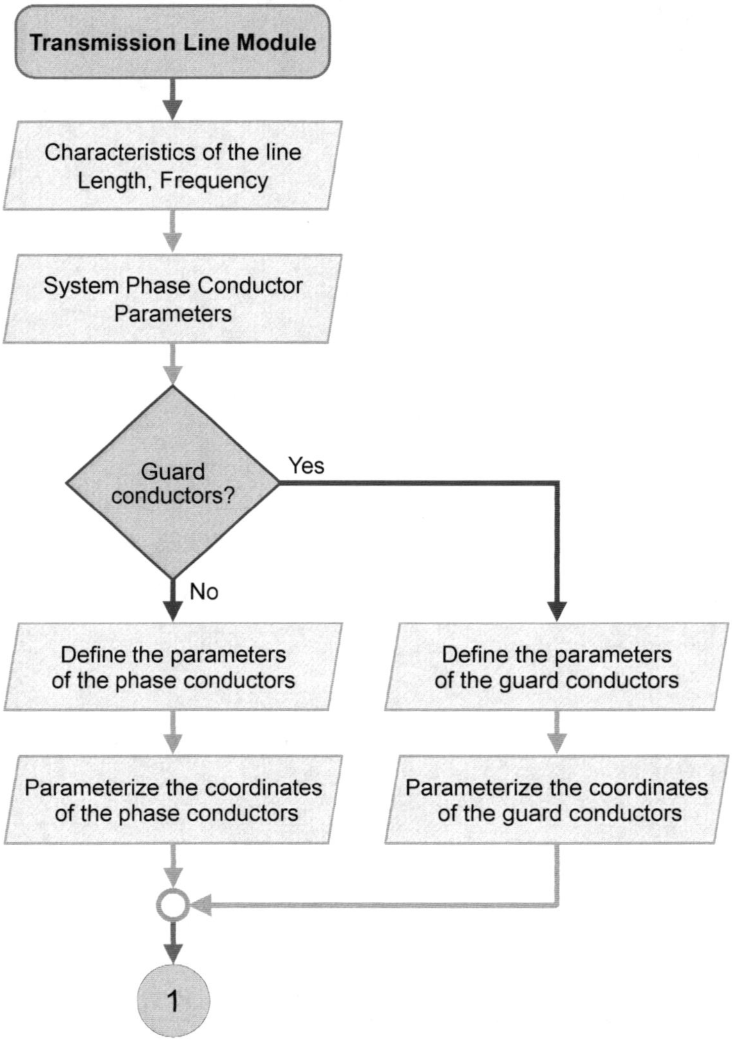

Figure 1. Flow chart for the characteristics of the line model

Figure 2 shows the mathematical process followed by the computational tool to determine the model in sequence components of a transposed transmission line. Additionally, a decision block can be observed that will allow showing in a later section the mathematical process followed by the computational tool to determine the model in sequence components of a non-transposed transmission line. It is relevant to note that once the program has calculated the model of the line, it can export the results in .xlsx or .txt files and to visualize the equivalent circuit for zero, positive and negative sequences.

Figure 3 reference source not found shows the mathematical process followed by the computational tool to determine the model in sequence components of a non-transposed transmission line, and for a transposed line, the tool allows exporting the results in .xlsx or txt format files and visualizing the equivalent circuit for zero, positive and negative sequences.

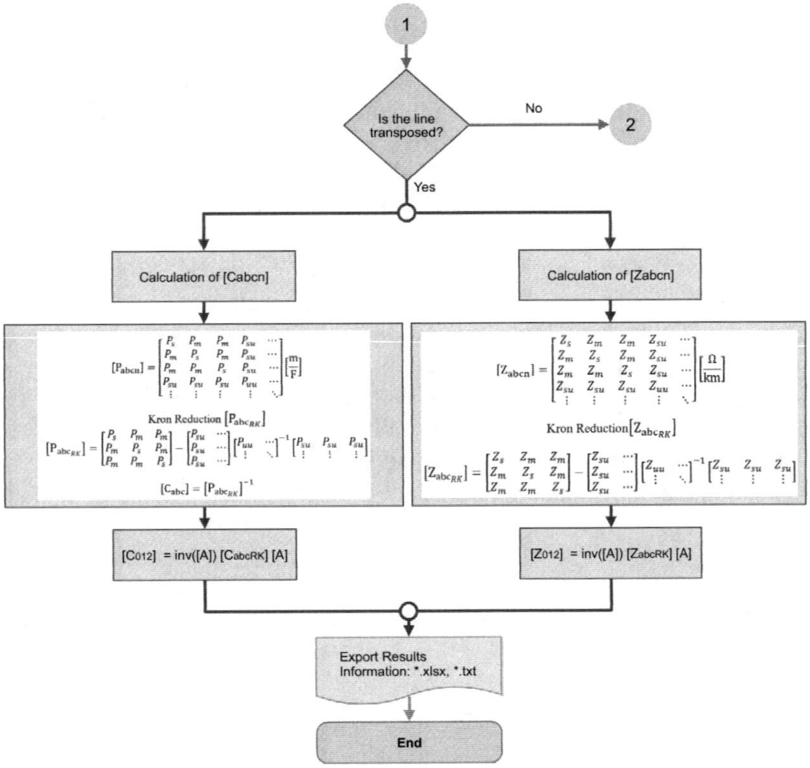

Figure 2. Flow chart for the transposed transmission lines model

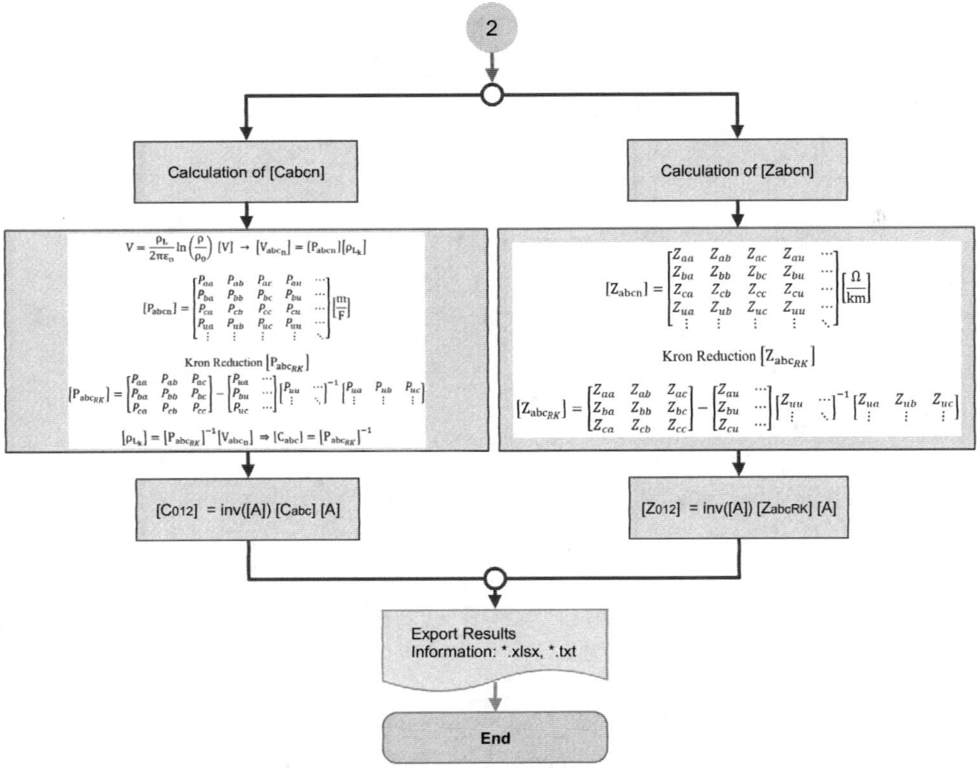

Figure 3. Flow chart for the non-transposed transmission lines model

3.2. Modular structure

The computational tool was developed in MATLAB R2020b App Designer. It consists of two modules that allow parameterizing the input variables and obtaining the model in sequence components, for transposed and non-transposed transmission lines.

3.2.1. Module 1: *Transmission Line Characteristics and Geometry*

This module is presented as the first graphical window that the user will see when running the computational tool. In this space, the user must configure the operating frequency and the characteristics of the line (length, sag, number of conductors per beam, and distance between the conductors of the beam). Subsequently, the user will have to configure the parameters of the phase and guard conductors, if the effect of the latter is considered, for this the user will have the possibility to choose between entering the information related to (RMG and Electrical Resistance at 75°C) or choosing a conductor from the manufacturer's library available in the tool. Finally, the user will have to configure the line geometry according to the adopted reference frame. Figure 4 shows the block diagram of the Transmission Line Characteristics and Geometry module.

This module is presented as the second graphical window that the user will see when running the computational tool. In this space, the type of line (transposed or not transposed) and the soil resistivity must be configured, for the latter the user will have the possibility to choose between the different types of soil available in the tool or to configure the value directly. Finally, the user will be able to choose the units in which wants the results to be presented, and clicking on the calculate button will show the results for Z_{abc}, C_{abc}, Z_{012}, C_{012} and the equivalent circuit for the zero, positive and negative sequences. Additionally, the panel to export the data in .xlsx or .txt formats will be activated. Figure 5 shows the block diagram of the Results module.

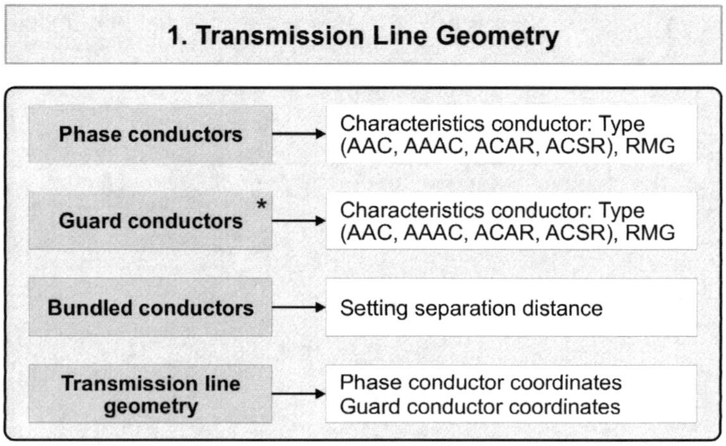

Figure 4. Structure of module 1 Characteristics and geometry of the transmission line.

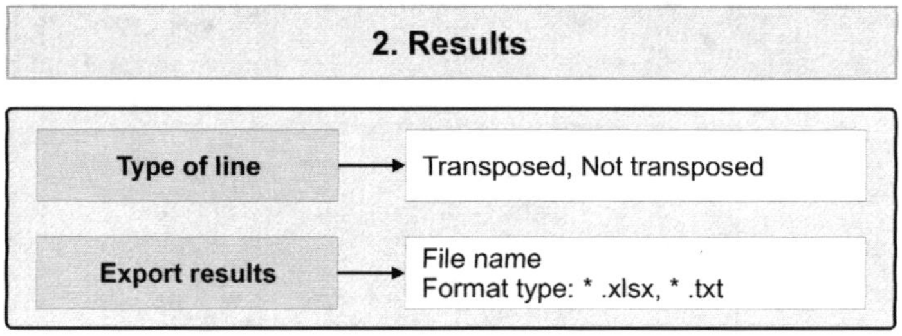

Figure 5. Structure of module 2 Results

4. Results

This section aims to evaluate the accuracy of the computational tool developed in Matlab App Designer. Therefore, two case studies are presented, and through a comparative analysis with the NEPLAN tool, the accuracy of the developed application is evaluated.

4.1. Graphic interface

Figure 6 shows the graphical interface of the computational tool developed in Matlab App Designer for a particular case study where the different graphical windows that the user will have to configure to obtain the model in sequence components of the transmission line of interest can be observed.

Figure 6. Graphical interface of the computational tool developed in MATLAB App Designer

4.2. Case studies

Two case studies are presented below; the first one corresponds to the model of a transposed transmission line, and the second one to a non-transposed line.

4.2.1. Case 1: Transposed transmission line

There is a transmission line with the configuration shown in Figure 2, which operates at a frequency of 60 [Hz]. The line is composed of 3 bundle conductors and two guard conductors. Table 1 shows the characteristics of the phase conductors, and Table 2 shows the guard conductors' characteristics. For this exercise, the length will be 1 [km] and the average resistivity 100 [Ω-m], which corresponds to a ground with a wet soil type (medium soil).

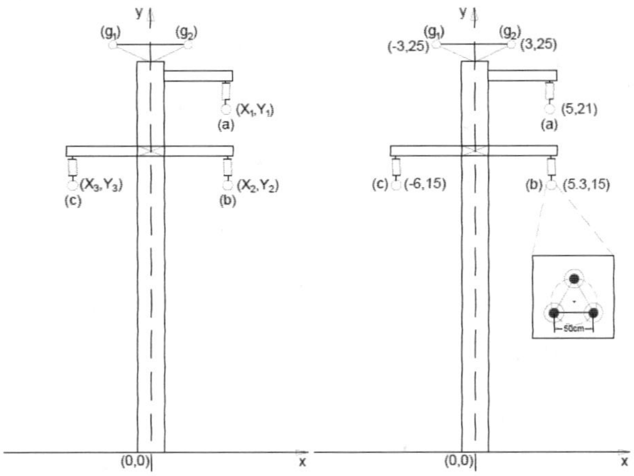

Figure 7. Transmission Line Case Study 1 (3 Conductor per bundle configuration)

Table 1. Characteristics of phase conductors (Case study 1)

Phase Conductors				
Cable Type	**Code**	**AWG / kcmil**	**RMG** [mm]	**Electrical Resistance 75°C** $\left[\frac{\Omega}{km}\right]$
ACSR (6/1)	Sparrow	2	2,58	1,10

Table 2. Characteristics of guard conductors (Case study 1)

Guard conductors				
Cable Type	**Code**	**AWG / kcmil**	**RMG** [mm]	**Electrical Resistance 75°C** $\left[\frac{\Omega}{km}\right]$
AAC (7)	Aster	2/0	3,81	0,521

Using the computational tool developed in Matlab App Designer, the sequence component model and the equivalent circuit for the transmission line shown in Figure X are determined. The results obtained by the computational tool are compared with the results provided by NEPLAN.

- Sequence Component Matrix Z_{012} (NEPLAN)

$$Z_{012} = \begin{bmatrix} 0{,}59847 + j0{,}88511 & 0 & 0 \\ 0 & 0{,}36667 + j0{,}35418 & 0 \\ 0 & 0 & 0{,}36667 + j0{,}35418 \end{bmatrix} \left[\frac{\Omega}{km}\right] \quad (25)$$

- Matrix of sequence components Z_{012} (MATLAB Computational Tool)

$$Z_{012} = \begin{bmatrix} 0{,}59847 + j0{,}88511 & 0 & 0 \\ 0 & 0{,}36667 + j0{,}35418 & 0 \\ 0 & 0 & 0{,}36667 + j0{,}35418 \end{bmatrix} \left[\frac{\Omega}{km}\right] \quad (26)$$

- Capacitance matrix in sequence components C_{012} (NEPLAN)

$$C_{012} = \begin{bmatrix} 7{,}49870 & 0 & 0 \\ 0 & 11{,}96924 & 0 \\ 0 & 0 & 11{,}96924 \end{bmatrix} \left[\frac{nF}{km}\right] \quad (27)$$

- Capacitance matrix in sequence components C_{012} (MATLAB Computational Tool)

$$C_{012} = \begin{bmatrix} 7{,}511 & 0 & 0 \\ 0 & 11{,}9611 & 0 \\ 0 & 0 & 11{,}9611 \end{bmatrix} \left[\frac{nF}{km}\right] \quad (28)$$

Table 3 shows the calculation of the percentage errors calculated for $\mathbf{Z_{012}}$ and $\mathbf{C_{012}}$

Table 3. Calculation of percentage errors of Z012 and C012 (Case of study 1)

	NEPLAN	**MATLAB Computational Tool**	**Error (%)**
Z_0	0,59847 + j0,88511	0,59847 + j0,88511	0
Z_1	0,36667 + j0,35418	0,36667 + j0,35418	0
Z_2	0,36667 + j0,35418	0,36667 + j0,35418	0
C_0	7,49870	7,511	0,16402
C_1	11,96924	11,9611	0,06800

	NEPLAN	MATLAB Computational Tool	Error (%)
C_2	11,96924	11,9611	0,06800

Figures 8, 9, and 10 display the equivalent circuit for the zero, positive, and negative sequences provided by the computational tool developed in Matlab App Designer.

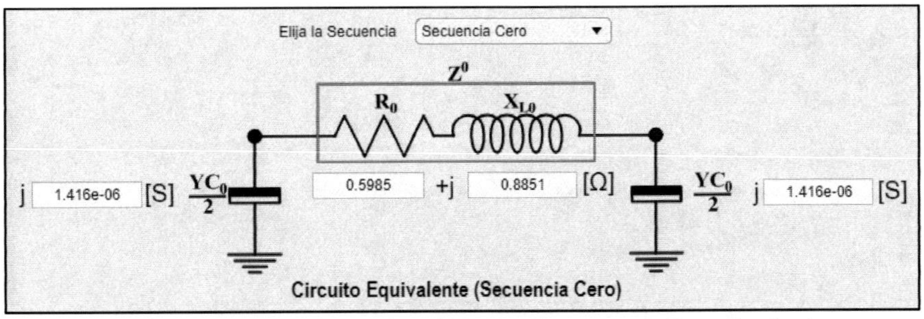

Figure 8. Equivalent Circuit (Zero Sequence) Case Study 1

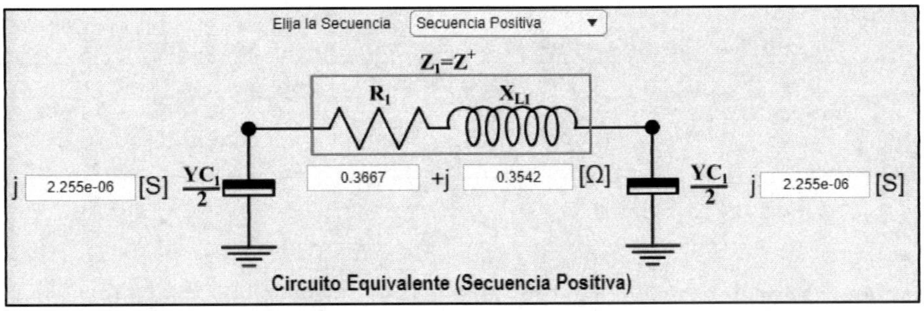

Figure 9. Equivalent Circuit (Positive Sequence) Case Study 1

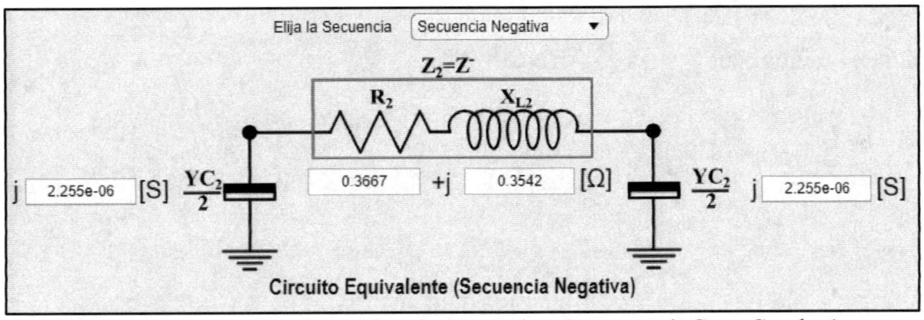

Figure 10. Equivalent Circuit (Negative Sequence) Case Study 1

4.2.2. Case 2: Non-transposed transmission line

There is a transmission line with the configuration shown in Figure 11, which operates at a frequency of 60 [Hz]. Table 4 shows the characteristics of the phase conductors. For this exercise, the length will be 1[km] and the average resistivity 100[Ω-m], which corresponds to a ground with a wet soil type (medium soil).

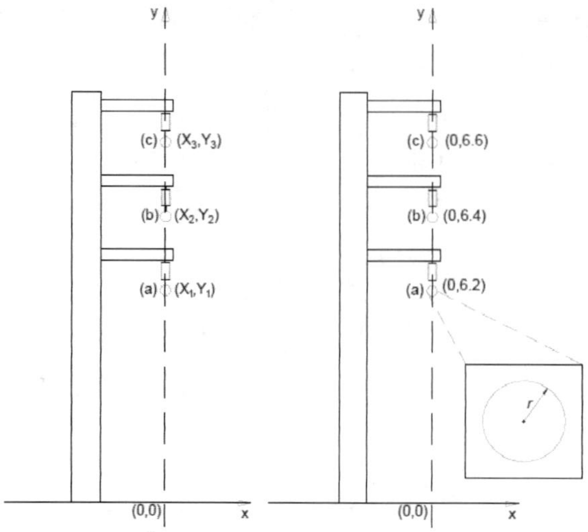

Figure 11. Transmission line case study 2

Table 4. Characteristics of phase conductors (Case Study 2)

Phase Conductors				
Cable Type	**Code**	**AWG / kcmil**	**RMG** [mm]	**Electrical Resistance 75°C** $\left[\frac{\Omega}{km}\right]$
ACSR (30/7)	Eagle	556,5	10,00	0,121

Using the computational tool developed in Matlab App Designer, the sequence component model and the equivalent circuit for the transmission line shown in Figure 11 is determined. The results obtained by the computational tool are compared with the results provided by NEPLAN.

- Sequence Component Matrix Z_{012} (NEPLAN 5.3.3)

$$Z_{012} = \begin{bmatrix} 0{,}29865 + j1{,}73358 & 0{,}01509 - j0{,}00871 & -0{,}01509 - j0{,}00871 \\ -0{,}01509 - j0{,}00871 & 0{,}12100 + j0{,}41690 & -0{,}03017 + j0{,}01742 \\ 0{,}01509 - j0{,}00871 & 0{,}03017 + j0{,}01742 & 0{,}12100 + j0{,}41690 \end{bmatrix} \left[\frac{\Omega}{km}\right] \tag{29}$$

- Matrix of sequence components Z_{012} (MATLAB Computational Tool)

$$Z_{012} = \begin{bmatrix} 0{,}29865 + j1{,}7336 & 0{,}015087 - j0{,}0087103 & -0{,}015087 - j0{,}0087103 \\ -0{,}015087 - j0{,}0087103 & 0{,}121 + j0{,}4169 & -0{,}030174 + j0{,}017421 \\ 0{,}015087 - j0{,}0087103 & 0{,}030174 + j0{,}017421 & 0{,}121 + j0{,}4169 \end{bmatrix} \left[\frac{\Omega}{km}\right] \tag{30}$$

- Capacitance matrix in sequence components C_{012} (NEPLAN 5.3.3)

$$C_{012} = \begin{bmatrix} 6{,}35959 & 0{,}60830 & 0{,}60830 \\ 0{,}60830 & 10{,}31692 & -0{,}27508 \\ 0{,}60830 & -0{,}27508 & 10{,}31692 \end{bmatrix} \left[\frac{nF}{km}\right] \tag{31}$$

- Capacitance matrix in sequence components C_{012} (MATLAB Computational Tool)

$$C_{012} = \begin{bmatrix} 6{,}3553 & 0{,}60789 & 0{,}60789 \\ 0{,}60789 & 10{,}3099 & -0{,}2749 \\ 0{,}60789 & -0{,}2749 & 10{,}3099 \end{bmatrix} \left[\frac{nF}{km}\right] \tag{32}$$

Table 5 shows the calculation of the percentage errors calculated for $\mathbf{Z_{012}}$ y $\mathbf{C_{012}}$

Table 5. Calculation of percentage errors of Z012 and C012 (Case of study 2)

	NEPLAN	MATLAB Computational Tool	Error (%)
Z_0	$0,29865 + j1,73358$	$0,29865 + j1,7336$	0
Z_1	$0,12100 + j0,41690$	$0,121 + j0,4169$	0
Z_2	$0,12100 + j0,41690$	$0,121 + j0,4169$	0
C_0	6,35959	6,3553	0,06745
C_1	10,31692	10,3099	0,06804
C_2	10,31692	10,3099	0,06804

It is important to note that in order to construct the equivalent in sequence components and the respective equivalent circuit, for a non-transposed line, the effect of the off-diagonal positions is ignored, and only the power of the diagonal positions is considered as in the transposed line.

Figures 12, 13, and 14 show the equivalent circuit for the zero, positive, and negative sequences provided by the computational tool developed in Matlab App Designer.

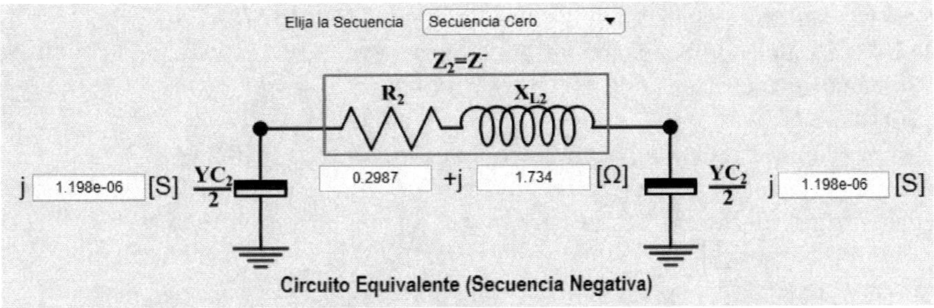

Figure 12. Equivalent Circuit (Zero Sequence) Case Study 2

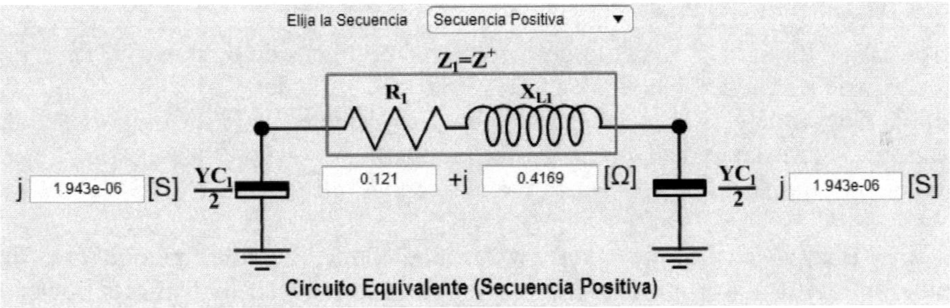

Figure 13. Equivalent Circuit (Positive Sequence) Case Study 2

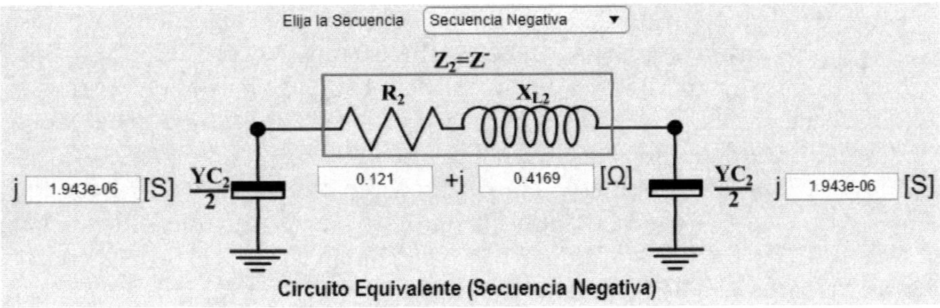

Figure 14. Equivalent Circuit (Negative Sequence) Case Study 2

5. Conclusions

In order to obtain the models in sequence components of the two case studies, computational tools such as Matlab and NEPLAN were used. The obtained results present practically negligible variations if it is considered approximations to 5 or 6 decimal places in calculations such as the serial model of the line. But the capacitance calculations present somewhat more significant variations. However, the data is closed between one software and the other.

The computational tool developed in Matlab allows obtaining the matrices in sequence components and the equivalent circuit for any transmission line regardless of its geometry, as long as it has a maximum of 2 guard wires. It also allows exporting the data obtained in Excel or TXT formats, which differs from some tools for power system analysis such as NEPLAN.

References

[1] Lv J. Transient stability assessment in large-scale power systems using sparse logistic classifiers. Int J Electr Power Energy Syst [Internet]. 2022;136:107626. Available from: https://www.sciencedirect.com/science/article/pii/S0142061521008589

[2] Li J, Yu T, Zhang X. Coordinated automatic generation control of interconnected power system with imitation guided exploration multi-agent deep reinforcement learning. Int J Electr Power Energy Syst [Internet]. 2022;136:107471. Available from: https://www.sciencedirect.com/science/article/pii/S0142061521007109

[3] Moradi R-A, Zeinali Davarani R. Introducing a new index to investigate voltage stability of power systems under actual operating conditions. Int J Electr Power Energy Syst [Internet]. 2022;136:107637. Available from: https://www.sciencedirect.com/science/article/pii/S0142061521008693

[4] Stracqualursi E, Araneo R, Faria JB, Andreotti A. Protection of distribution overhead power lines against direct lightning strokes by means of underbuilt ground wires. Electr Power Syst Res [Internet]. 2022;202:107571. Available from: https://www.sciencedirect.com/science/article/pii/S0378779621005526

[5] Kurokawa S, Daltin RS, Prado AJ, Pissolato J. An alternative modal representation of a symmetrical nontransposed three-phase transmission line. IEEE Trans Power Syst. 2007;22(1):500–1.

[6] Calderon EY, Quiroga DE. Cálculo de parámetros en líneas de transmisión de energía eléctrica aéreas en corriente alterna. Universidad Industrial de Santander; 2017.

[7] Mustafa T, Cabral SHL, Almaguer HAD, Meyer LH, Puchale LHB, Cereja JE, et al. A practical application of an analytical method for modeling power transmission lines. In: 2017 International Symposium on Electromagnetic Compatibility-EMC EUROPE. IEEE; 2017. p. 1–4.

[8] Gaur VK, Bhalja BR, Saber A. New ground fault location method for three-terminal transmission line using unsynchronized current measurements. Int J Electr Power Energy Syst [Internet]. 2022;135:107513. Available from: https://www.sciencedirect.com/science/article/pii/S0142061521007523

[9] Liu Y, Wang B, Zheng X, Lu D, Fu M, Tai N. Fault location algorithm for non-homogeneous transmission lines considering line asymmetry. IEEE Trans Power Deliv. 2020;35(5):2425–37.

[10] Niu G, Xiao T, Pei W, Zhu M, Yan Z, Zhou L, et al. Fault analysis of power transmission line in a generalized state-space model perspective. In: IECON 2017-43rd Annual Conference of the IEEE Industrial Electronics Society. IEEE; 2017. p. 239–43.

[11] Henke HTJ, Gardumi F, Howells M. The open source electricity Model Base for Europe - An engagement framework for open and transparent European energy modelling. Energy [Internet]. 2022;239:121973. Available from: https://www.sciencedirect.com/science/article/pii/S0360544221022210

[12] Andreev M, Suvorov A, Askarov A, Rudnik V, Bhalja BR. Novel approach for relays tuning using detailed mathematical model of electric power system. Int J Electr Power Energy Syst [Internet]. 2022;135:107572. Available from: https://www.sciencedirect.com/science/article/pii/S0142061521008061

Journal of Physics: Conference Series **2135** (2021) 012008 doi:10.1088/1742-6596/2135/1/012008

[13] Correa J, Chacon J. Campo eléctrico inducido en objetos cercanos a una línea de transmisión de alta tensión. Universidad Industrial de Santander; 2017.
[14] Bayliss CR, Bayliss C, Hardy B. Transmission and distribution electrical engineering. Elsevier; 2012.
[15] Gonen T. Modern power system analysis. CRC Press; 2013.
[16] Das JC. Understanding symmetrical components for power system modeling. John Wiley & Sons; 2016.

Optimal Power Dispatch of Dispersed Sources in Direct-Current Networks with Nonlinear Loads

O. D. Montoya[1,2], **V. M. Garrido**[3] **and L. F. Grisales-Noreña**[4],

[1]Facultad de Ingeniería, Universidad Distriral Francisco José de Caldas, Bogotá, Colombia.
[2]Laboratorio Inteligente de Energía, Universidad Tecnológica de Bolívar, Cartagena, Colombia.
[3]Facultad de Ingenierías y Arquitectura, Universidad de Pamplona, Pamplona, Colombia.
[4]Departamento de Electromecánica y Mecatrónica, Instituto Tecnológico Metropolitano, Medellín, Colombia.

E-mail: odmontoyag@udistrital.edu.co

Abstract. The problem of the optimal power dispatch of dispersed generators in direct-current networks under the presence of nonlinear loads (constant power terminals) is addressed through a combinatorial optimization strategy by using a master-slave solution methodology. The optimal power generation in the dispersed is solved in the master optimization stage through the application of the vortex-search algorithm. Each combination of the power outputs at the dispersed generation sources is provided to a power flow methodology known as the hyperbolic power flow approach for direct current networks. The main advantage of the proposed optimization method corresponds to the possibility of solving a complex nonlinear programming problem via sequential quadratic programming, which can be easily implemented at any programming language with low computational effort and high-quality results. The computational tests of the master-slave optimization proposal are evaluated in a 21-bus system, and the numerical results are compared with the implementation of the exact nonlinear programming model in the General Algebraic Modeling System (i.e., GAMS). All the computational results are conducted through the MATLAB programming environment licensed by Universidad Tecnológica de Pereira for academic usage.

1. Introduction

Nowadays, direct current (DC) microgrids are a powerful electrical network tendency that becomes a possible economic and reliable alternative for providing electrical services to millions of users around the world [1]. The main advantage of using DC grids lie in the possibility of integrating a lot of dispersed energy resources, such as renewable generation [3] (solar, wind technologies, among others), and energy storage systems [2] (batteries, supercapacitors or fuel-cells) with DC-DC converters or ac-DC inverters, thus reducing the back-to-back topologies required to integrate them into conventional AC power grids [4]. Those reductions must be reflected in the total cost of the dispersed energy resources, as well as lower power loss and high electrical efficiency of DC networks in comparison with their AC counterparts [5]. Another important driver of the popularization of DC grids is that they are easier to analyze and operate because disappear frequency and reactive power control, key concepts of classical AC grids [5].

One of the most important problems in DC networks is the optimal power flow problem (OPF) because it represents an essential tool for planning, operating, and controlling DC networks [5–7]. The OPF problem is interesting for the research community because it is nonlinear and non-convex, and solving it requires of purposing new and efficient methodologies [8, 9].

In the case of Colombian electrical power system, regulatory entities have impulsed the massive integration of renewable energy sources and energy storage devices in all the voltage levels; in addition, the concept of microgrids for isolated applications has also impulsed by utilities for improving their grids in terms of voltage profiles and power loss. These characteristics have converted DC power systems into a powerful alternative for improving the efficiency and reliability of the power system in Colombia, and this paper tries to contribute with powerful tools for its analysis as is the case of the OPF solvers.

According to the review of the state-of-the-art above, only two combinatorial optimization methods have been applied to solve OPF problems in DC power grids, such as the continuous genetic algorithm proposed in [1] and the black hole optimizer presented in [10]. In that sense, this paper identifies a research gap in the field and offers a VSA approach [11]. The main advantage of the proposed approach in this work, in comparison with convex methods, lies in the fact that the number of variables of the power flow problem remains constant and no eigenvalue decomposition is required to recover power flow variables [8].

To complement the proposed VSA algorithm in the master stage, a Taylor-based method recently proposed in [12] for power solution at the slave stage. This combination produces a new hybrid optimization approach ahead named VSA-TBM. Additionally, this method does not require the usage of specialized software, and it can be implemented in any programming language since its solution methodology is pure-algorithmic [1, 10].

The remainder of this paper is organized as follows: Section 2 presents the conventional OPF problem for DC power grids with a focus on the possibility of including dispersed generation in the grid by controlling their percentages of penetration. Section 3 introduces the main characteristics of the VSA method as well as its evolution process and its application to the master problem; in addition, it shows the conventional TBM formulation for solving power flow problems in DC power grids with its use to the slave problem. Section 4 describes the test system, the numerical results, as well as the comparative methods. Finally, Section 5 draws the main conclusions and possible future research derived from this work, followed by acknowledgments and references.

2. Mathematical model

The OPF problem in electrical distribution networks operated with DC technology can be represented as a nonlinear non-convex optimization problem, where the most common objective function analyzed in literature is the minimization of the total grid power losses. This objective function is constrained with the power equilibrium at each node, the voltage regulation bounds, and the maximum capabilities of power generation, among others. The most complicate constraint is the hyperbolic relation among voltages and powers in the power balance constraints [8, 9]. The entire OPF model is listed below.

Objective function

$$\min p_{loss} = \sum_{i=1}^{n}\sum_{j=1}^{n} g_{ij} v_i v_j, \tag{1}$$

where p_{loss} is the performance indicator (objective function) regarding the total power losses in all the distribution lines of the DC network (i.e., power losses in the resistive effect at each conductor); g_{ij} is the conductance component that connects nodes i and j, which is obtained from the nodal conductance matrix; v_i and v_j are the values of the voltage at nodes i and j, respectively. It is worth mentioning that n represent the total number of nodes of the electrical network under analysis.

Set of constraints:

$$p_i^s = v_i \sum_{j=1}^{n} g_{ij} v_i v_j, \; \forall i \in \mathcal{S}, \tag{2}$$

$$p_i^{dg} - p_i^d = v_i \sum_{j=1}^{n} g_{ij} v_i v_j, \; \forall i \in \mathcal{N} - \mathcal{S}, \tag{3}$$

$$p_i^s \geq 0, \ \forall j \in \mathcal{S}, \quad (4)$$

$$v_i = v_{is}, \ \forall i \in \mathcal{S}, \quad (5)$$

$$0 \leq p_i^{dg} \leq p_i^{dg,\max}, \ \forall i \in \mathcal{N} - \mathcal{S}, \quad (6)$$

$$v_i^{\min} \leq v_i \leq v_i^{\max}, \ \forall i \in \mathcal{N} - \mathcal{S}, \quad (7)$$

where p_i^s is a vector that contains all the power injection in the slack sources; p_i^{dg} is a vector that contains all the power injections in the dispersed generators; p_i^d represents the vector that contains all the power consumptions in the constant power nodes; $p_i^{dg,\max}$ is the maximum power generation bound in the dispersed source connected at node i; $v_i^{\min}$ and $v_i^{\max}$ represent the minimum and maximum voltage limits associated at each node i contained in the distribution grid under study. Note that $\mathcal{S}$ is the set that contains all the slack sources.

The interpretation of the OPF model given from Equations (1) to (7) is the following: in Equation (1) is presented the performance indicator of the studied OPF problem which consists in the minimization of the total grid losses minimization; in Equations (2) and (3) are defined power equilibria constraints of the network, which are separated for the slack and demands nodes, respectively. Equations (4) and (5) define the nature of the decision variables, i.e., that the power generations must be positive definite and the constant magnitudes of the voltage profile at each slack node. On the other hand, Equations (6) and (7) define the positiveness characteristic of the power injected by dispersed sources as well as that the voltages must be contained between two positive limits known as the voltage regulation bounds

Owing to the nonlinear non-convexity nature of the OPF problem under analysis, in the current literature can be found two main solution ways that allow reaching high-quality solutions with low computational effort. These alternatives are: *i)* use convex programming approximations [8, 13], or *ii)* apply combinatorial optimization methods based on metaheuristics [1, 10]. In this research, we adopt the second way to solve the OPF problem as will be detailed in the next section.

3. Strategy of solution

To address the OPF problem in electrical DC distribution networks here is proposed a combinatorial optimization approach based on master-slave strategy; where the master stage is guided by the vortex-search algorithm, and the slave strategy corresponds to a power flow based on the hyperbolic formulation adapted for DC grids.

3.1. Master stage

To solve the OPF problem in DC distribution networks, we proposed a master strategy based on VSA, which takes upon of defining the optimal power injections in all the dispersed generators. The general steps for implementing the VSA to solve the OPF problem are listed in Algorithm 1.

To know about all the mathematical elements and symbols presented in Algorithm 1, the following references are useful [11] and [14].

3.2. Slave stage

Power flow analysis for DC grids can be analyzed with different methodologies, including successive approximations, Gauss-Seidel, Newton-Raphson, and Taylor-based methods (TBMs), among others. Here, we select a TBM recently reported in [12]. This approach works by an iterative procedure based on a simplified Jacobian rule as presented in the recursive formula defined by Equation (8).

$$v_d^{t+1} = \left[G_{dd} + \mathbf{diag}^{-2}(v_d^t)\mathbf{diag}\left(p^{dg} - p^d\right)\right]^{-1} \left[2\mathbf{diag}^{-1}(v_d^t)\left[p^{dg} - p^d\right] - G_{ds}v_s\right], \quad (8)$$

where p^{dg} is a vector that contains all the power inputs in the dispersed sources, p^d is the vector that contains all the powers absorbed in the constant power terminals; v_d is a vector that includes all the

Data: Select the DC grid topology and define all the parameters of the VSA
Define the starting center of the hypersphere μ_m;
Define the initial radius of the hypersphere r_m;
Generate the initial population S^m;
Verify the lower and upper bounds of all the individuals.;
Evaluate each s_i^m and find s_{best}^m;
for $m = 1 : m_{\max}$ **do**
  Update the center of the hypersphere μ_{m+1};
  Calculate the new radius of the hypersphere r_{m+1};
  Calculate the new popilation S^{m+1};
  Verify the feasibility of all the solution individuals;
  Verify the lower and upper bounds of all the individuals.;
  Evaluate all the individuals s_i^{m+1} and find s_{best}^{m+1};
  Revise the non-consecutive improvements of the objective function;
  if $k \geq k_{\max}$ **then**
    Select as solution of the problem μ_{m+1};
    Return the solution of the OPF problem;
    break;
  end
end
Result: Provide the solution of the OPF problem

Algorithm 1: General application of the VSA to solve the OPF problem in DC networks

voltage variables in the constant power terminals; v_s represents a vector that includes all the voltage in the slack sources (i.e., a vector of known variables); and G_{ds} and G_{dd} are components of the conductance matrix that connects slack and demands nodes among them. In addition, t is incremented until $t_{\max}$ (maximum number of iterations) and the search stop when $\max\left\{\left|v_d^{t+1} - v_d^t\right|\right\} \leq \epsilon$. It is important to mention that ϵ represents the maximum convergence error allowed between two consecutive iterations. In specialized literature, this parameter is typically selected as 1×10^{-10} [12].

4. Numerical validation

The computational validation of the proposed master-slave optimization method for OPF solution in DC distribution networks is made in the 21-node test feeder [8]. This test system consists of 21 nodes and 20 lines with multiple constant power terminals. In addition, said system has a unique voltage controlled node (i.e., substation bus at node 1). All of the information of this test system can be consulted in [8]. This study analyzes the possibility of installing from three dispersed generators considering penetration percentages from 20% to 60% of the total power consumption. Note that, via heuristic search methods, nodes 9, 12 and 16 were selected for DGs' location [10].

Table 1 shows the generation at each DG considering different percentages of power penetration, total power generation, and objective function values that are total power losses in the network DC. Note that, in each simulation case, the maximum power injection allowed into the DC grid reaches 1,0954 p.u, 2.2160 p.u and 3.3240 p.u for dispersed generator penetration scenarios of 20%, 40% and 60%, respectively. In that sense, from the fifth column in Table 1, it can be noticed that the proposed VSA-TBM method as well as the GAMS optimizer use more than 99% of the maximum generation allowed to reduce total power losses in the DC network. In terms of objective function minimization, note that the method proposed in this work presents an estimation error lower than 0.30% (compared to the SCIP solver) for each simulation case. Therefore, such hybrid VSA-TBM method offers adequate numerical convergence compared to large-scale nonlinear packages.

It is important to stand out that the power delivered by each dispersed generator exhibits small

Table 1: Numerical comparison between the proposed approach and the GAMS package [15]

Solutions provided by the VSA–TBM method proposed in this work						
Penetration [%]	DG 9 [p.u]	DG 12 [p.u]	DG 16 [p.u]	Total Gen [p.u]	Losses [p.u]	Reduction [%]
20%	0.0168	0.0405	1.0381	1.0954	0.13823	49.91
40%	0.2472	0.6981	1.2707	2.2160	0.06611	76.04
60%	0.8451	1.0294	1.4495	3.3234	0.03061	88.90
Solutions provided by the GAMS and the SCIP solver						
20%	3×10^{-6}	0.1426	0.9654	1.1080	0.13662	50.50
40%	0.2464	0.6970	1.2726	2.2160	0.06621	76.00
60%	0.8442	1.0254	1.4544	3.2400	0.03071	88.87

variations compared to GAMS optimizer results in the solutions obtained by means of the proposed approach. These differences can be attributed to the nonlinear continuous nature of the OPF problem and to the use of metaheuristic method, which implies that multiple dispersed generator power combinations with identical objective functions could exist. The proposed VSA-TBM method and the comparative method via GAMS implementation exhibit similar performance in terms of power losses reduction, which clearly validates the proposed approach in terms of numerical convergence. In addition, when the reduction of the final power loss respect to the case without dispersed generation are compared as can be seen in the seventh column of Table 1, we can be sure that our proposed approach is efficient and reliable for OPF analysis in relation to well-known commercial solvers.

5. Conclusion and future work

This paper presented a combination of the vortex search algorithm approach in conjunction with the Taylor series-based method for solving the OPF problem in DC power grids. The VSA method is a soft variant of the widely-known random search and pattern search algorithms, which offer an adequate trade-off between exploration and exploitation of the solution space for continuous optimization models as the case of the OPF problem. To confirm this advantage, the large-scale nonlinear optimization package SCIP solver (available for GAMS) was used as comparative method, which allowed observing the excellent numerical convergence of the proposed hybrid VSA-TBM method for OPF analysis.

As future works, the VSA-TBM method presented in this paper could be used for solving the problem of optimal location and sizing of dispersed generation in DC power grids in conjunction with binary metaheuristic techniques or use it within an economic dispatch strategy of DC networks.

References

[1] Montoya O D, Gil-González W, and Grisales-Noreña L F 2018 Optimal Power Dispatch of DGs in DC Power Grids: a Hybrid Gauss-Seidel-Genetic-Algorithm Methodology for Solving the OPF Problem. *WSEAS Transactions on Power Systems*, 13(13) pp 335–346.

[2] Patterson M, Macia N F, and Kannan A M 2015 Hybrid microgrid model based on solar photovoltaic battery fuel cell system for intermittent load applications. *IEEE Transactions on Energy Conversion*, 30(1) pp 359–366.

[3] Ellabban O, Abu-Rub H, and Blaabjerg F 2014 Renewable energy resources: Current status, future prospects and their enabling technology. *Renewable Sustainable Energy Rev.*, 39 pp 748–764.

[4] Parhizi S, Lotfi H, Khodaei A, and Bahramirad S 2015 State of the art in research on microgrids: A review. *IEEE Access*, 3 pp 890–925..

[5] Garces A 2017 Uniqueness of the power flow solutions in low voltage direct current grids. *Electr. Power Syst. Res.*, 151(Supplement C) pp 149–153.

[6] Li J, Liu F, Wang Z, Low S H, and Mei S 2018 Optimal Power Flow in Stand-Alone DC Microgrids. *IEEE Trans. Power Syst.*, 33(5) pp 5496–5506.

[7] Garces A 2018 On Convergence of Newtons Method in Power Flow Study for DC Microgrids. *IEEE Trans. Power Syst.*, 33(5) pp 5770–5777.

[8] Montoya O D, Gil-González W, and Garces A 2019 Optimal Power Flow on DC Microgrids: A Quadratic Convex Approximation. *IEEE Trans. Circuits Syst. II*, 66(6) pp 1018–1022

[9] Simpson-Porco J W, Dorfler F, and Bullo F 2015 On resistive networks of constant-power devices. *IEEE Transactions on Circuits and Systems II: Express Briefs*, 62(8) pp 811–815.
[10] Velasquez O S, Montoya O D, Garrido-Arevalo V M, and Grisales-Noreña L F 2019 Optimal Power Flow in Direct-Current Power Grids via Black Hole Optimization. *Advances in Electrical and Electronic Engineering*, 17(1) pp 24–32.
[11] Aydin O, Tezcan S S, Eke I, and Taplamacioglu M C 2017 Solving the Optimal Power Flow Quadratic Cost Functions using Vortex Search Algorithm. *IFAC-PapersOnLine*, 50(1) pp 239–244 20th IFAC World Congress.
[12] Montoya O D, Garrido V M, Gil-González W, and Grisales-Noreña L F 2019 Power Flow Analysis in DC Grids: Two Alternative Numerical Methods. *IEEE Trans. Circuits Syst. II*, pp 1–5.
[13] Montoya O D, Gil-González W, and Garces A 2019 Sequential quadratic programming models for solving the OPF problem in DC grids. *Electr. Power Syst. Res.*, 169 pp 18–23.
[14] Dogan B and Olmez T 2015 Vortex search algorithm for the analog active filter component selection problem. *AEU - International Journal of Electronics and Communications*, 69(9) pp 1243–1253.
[15] GAMS Development Corp. GAMS free demo version [Online:] `https://www.gams.com/`.

Black hole optimizer for the optimal power injection in distribution networks using DG

Oscar Danilo Montoya*, Diego Armando Giral-Ramírez, Luis Fernando Grisales-Noreña*****

*Universidad Distrital Francisco José de Caldas. Colombia. odmontoyag@udistrital.edu.co
*Universidad Tecnológica de Bolívar. Colombia. omontoya@utb.edu.co
**Universidad Distrital Francisco José de Caldas. Colombia. dagiralr@udistrital.edu.co
***Instituto Tecnológico Metropolitano. Medellín. Colombia. luisgrisales@itm.edu.co

Abstract. The optimal sizing of Distributed Generators (DG) in electric power distribution networks is carried out through a metaheuristic optimization strategy. To size DG it is proposed an optimal power flow model is formulated by considering that the location of these sources has been previously defined by the distribution company. The solution of the optimal power flow is reached with the Black Hole Optimizer (BHO). A methodology is used master-slave optimization methodology, where the BHO (i.e., master stage) defines the sizes of the DG and the slave stage evaluates the objective function with a load flow algorithm, this work using the triangular-based power flow method. Numerical results in the 33-node and the 69-node test system demonstrates the effectiveness and robustness of the proposed approach when compared with literature results.

1. Introduction

The main characteristic of electrical distribution networks is associated with their construction, since these are typically designed in radial form to minimize investment costs in conductors [1], and also for simplifying the coordination of the protective devices [2]. These features in the electrical distribution networks make these highly affected by energy losses inherent to the distribution tanks, since part of the purchase energy at the substation node is transformed into heat energy [3]. The problem of power losses in electrical distribution networks in the specialized literature has been proposed multiple compensation strategies that include: (i) installation of capacitor banks and distribution system compensators (D-STATCOM) [4]; (ii) installation of battery energy storage systems [5]; (iii) grid reconfiguration [6]; and optimal placement and sizing of DG [7]. All of these approaches have positive effects regarding the amount of power losses reduction, being the optimal location and sizing of the DG the approach that allow reductions higher that 50 % [8].

In the literature have been proposed multiple optimization approaches that use exact and metaheuristic optimization approaches to solve this Mixed-Integer Nonlinear Programming (MINLP) model. Regarding exact approaches, [9] have presented a mixed-integer convex methodology that allows transforming the exact MINLP model into a mixed-integer second-order cone programming (MISOCP), that allows solving the location and sizing of DG using the combination of the branch & bound and the interior point method. The disadvantage of this approach is the usage of convex optimization packages that works with binary variables such as CVX and the GUROBI solver, which is no available in all the programming platforms. Authors of [10] have presented a combination of the sine-cosine algorithm in its discrete form with the conic formulation of the optimal power flow problem to solve the problem of optimally siting and sizing of DG in electrical distribution networks; however, the implementation also requires packages that solve SOCP models efficiently. In references [7, 11] and [12] have formulated MINLP models to locate and dimension DG and their solutions have been obtained using GAMS, the main advantage of these implementations is that the authors have concentrated their efforts to obtain accurate optimization models for representing the optimization problem; however, the GAMS package has a high probability of reporting local solutions due to the non-convex nature of the exact MINLP model. Regarding the combinatorial optimization, i.e., metaheuristics, in literature can be found multiple approaches to solve the optimal location and sizing DG, some of them are: krill-herd optimization algorithm [1, 13]; genetic algorithms [14, 15, 16, 17]; particle swarm optimization [3, 18, 19]; sunflower optimization algorithm [20]; population-based incremental learning [8]; tabu search algorithm [21, 22]; and flower pollination algorithm [23, 24], among others.

The main characteristic of the aforementioned metaheuristic approaches is that these work under a master-slave optimization approach [17], where the master-stage is entrusted with the optimal location of the DG, while the slave-stage set the optimum size of the GDs [8]. In the solution of the slave stage are typically used combinatorial optimization methods, being the particle swarm optimizer the most classical method in this stage [3, 16]. According to the literature review, for the problem of the optimal location and sizing of DG [10], the slave stage is the most sensitive part of the optimization algorithm, since the reduction of the power losses is completely dependent on the sizes of the DG; which implies that in this stage are require simple but efficient optimization techniques that ensure an adequate performance between the optimal solution and the processing times. According to what was previously written, this research proposed the application of the BHO for solving the slave stage (i.e., optimal power flow) in the problem of the optimal placement and sizing of DG in electrical distribution networks, by assuming that the distribution company provides the points where the DG will be located. The BHO is selected since in the current literature only the authors of [25] present the application of this method for the optimal dispatch of constant power sources for direct-current networks which are simplest when compared with AC electrical distribution grids addressed in this research.

In the current literature, the BHO has been proposed for solving different optimization problems, such as: data clustering [26]; set covering problem [27]; minimization of nonlinear and non-convex functions [28]; spam detection [29]; biclustering microarray gene expression data [30]; and optimal power flow solutions in DC networks [25, 31], among others. It is worth mentioning that the application of the BHO has no been previously reported in the specialized literature corresponds to the main contribution of this research to solve the problem of the optimal sizing of the DG in AC electrical distribution networks considering their locations perfectly known.

The article has six sections: Section 2 presents the mathematical formulation of the optimal location and optimal sizing of DG in AC electrical distribution grids in the complex domain, where the binary variables regarding the location of the generators are introduced for the completeness in the mathematical formulation. Section 3 presents the main characteristics of the BHO approach. This section also presents the power flow methodology used to determine the power losses once the optimal sizes of the DG are defined by the BHO. Section 4 presents the radial test systems used. Section 5 presents the numerical results of the proposed optimization approach and its comparisons with literature reports. Finally, Section 6 presents the conclusions of this research.

2. Mathematical formulation

The problem of the optimal sizing of DG in electrical distribution networks can be formulated with a MINLP model, where the binary part regarding the location of the distributed sources is assumed known; and the continuous part is associated with the resulting optimal power flow model, which allows determining the optimal sizes of the distributed sources. The complete optimization model, including the binary variables is formulated as follows.

2.1. Objective function
For the optimal sizing of DG is to the minimization of the amount of active power losses during the peak load condition, which is presented in Equation (1).

$$\min p_{\text{loss}} = \text{Real}\left\{\mathbb{V}^T \mathbb{Y}_{\text{bus}}^{\star} \mathbb{V}^{\star}\right\}, \tag{1}$$

Where p_{loss} is the amount of active power losses in all the branches of the network; $\mathbb{V} \in \mathcal{C}^{n\times 1}$ is the vector that contains the complex voltages in all the nodes of the network, being n the total nodes of the network; $\mathbb{Y}_{\text{bus}} \in \mathcal{C}^{n\times n}$ is the bus admittance matrix that has the information about the physical interconnection among nodes. Note that $\mathcal{C}$ represents the set of complex numbers.

2.2. Set of constraints
The problem of the optimal sizing of distributed sources in electrical distribution grids is typically constrained by power balance equations, the capacities of the generators and the voltage regulation bounds, among others (Equations (2) - (6)).

$$\mathbb{S}_{gc}^{\star} + \mathbb{S}_{dg}^{\star} - \mathbb{S}_{d}^{\star} = \textbf{diag}\left(\mathbb{V}^{\star}\right) \mathbb{Y}_{\text{bus}} \mathbb{V}, \tag{2}$$

$$\mathbf{S}_{gc}^{\min} \leq |\mathbb{S}_{gc}| \leq \mathbf{S}_{gc}^{\max}, \tag{3}$$

$$x \cdot \mathbf{S}_{gd}^{\min} \leq |\mathbb{S}_{dg}| \leq x \cdot \mathbf{S}_{gc}^{\max}, \tag{4}$$

$$\mathbf{V}^{\min} \leq |\mathbb{V}| \leq \mathbf{V}^{\max}, \tag{5}$$

$$\mathbf{1}^T x \leq N_{\max}^{gd}, \tag{6}$$

Where $\mathbb{S}_{gc} \in \mathcal{C}^{n\times 1}$, $\mathbb{S}_{gd} \in \mathcal{C}^{n\times 1}$, and $\mathbb{S}_{d} \in \mathcal{C}^{n\times 1}$ are the column vectors with all the associated variables associated with the conventional and DG, and the load parameters, respectively; $\mathbf{S}_{gc}^{\min} \in \mathcal{R}^{n\times 1}$ and $\mathbf{S}_{gc}^{\max} \in \mathcal{R}^{n\times 1}$ are the vectors that contain the lower and upper apparent power bounds associated with the conventional sources; $\mathbf{S}_{gd}^{\min} \in \mathcal{R}^{n\times 1}$ and $\mathbf{S}_{gd}^{\max} \in \mathcal{R}^{n\times 1}$ represent the minimum and maximum apparent power injection bounds regarding the DG that will be installed in the electrical distribution grid; x is the binary vector that decides in where nodes will be located or no a DG; $\mathbf{V}^{\min} \in \mathcal{R}^{n\times 1}$ and $\mathbf{V}^{\max} \in \mathcal{R}^{n\times 1}$ are the vectors that contain all the minimum and maximum voltage regulation bounds of the grid; $N_{\max}^{gd} \in \mathcal{R}$ is an scalar which represents with the maximum number of DG available for installation. $\mathcal{R}$ is the set that contains the real numbers; $w \cdot y$ represents the element-wise product between two vectors, i.e., w and y; $\textbf{diag}\,(w)$ is a mathematical operation that becomes the vector w in a diagonal matrix; and $\mathbf{1} \in \mathcal{R}^{n\times 1}$ is a column vector of ones. Additionally, is assumed that these operate with unity power factor, which implies that $\mathbb{S}_{dg} = \mathbf{P}_{dg}$.

2.3. Model interpretation
The MINLP model defined from (1) to (6) is an optimization model defined in the complex domain, where: Equation (1) defines the objective function associated with the minimizing of the active power losses in all the conductors of the network; Equation (2) represents the complex apparent power balance in all the nodes of the network, i.e., the application of the second Tellegen's theorem at each node of the network; Inequality constraints (3) and (4)

correspond to the restrictions associated with the minimum and maximum power generation limits in the conventional and DG; Inequality constraints (5) is known as the voltage regulation constraint which is entrusted with guaranteeing all the voltages in the distribution network in their admissible bounds. These limits are typically defined by the grid operator based on the applicable regulatory policies [11]. Finally, Inequality constraint (6) determines the maximum number of DG available for installation inside of the distribution network.

In this research is assumed that: (i) the location of the DG were previously defined by the grid operator, which implies that to compare numerical results regarding the power losses minimimization, the different locations provided in literature will be tested with our proposal optimization algorithm, and (ii) to solve the problem of the optimal sizing is proposed the application of a master-slave optimization strategy, where the master stage corresponds to the BHO which is entrusted with the determination of the DG sizes, and the slave stage is a power flow method which allows calculating the objective function value. The description of the optimization methodology will be presented in the next section.

3. Black hole optimizer

The BHO is a continuous metaheuristic optimization algorithm belonging the nature-inspired techniques that emulates the interaction among stars and black holes located in the center of the galaxies [25]. An exemple of this interaction is show in Fig. 1. The black optimization use the position of the stars at the black hole as potential solutions nonlinear programming models with multiple constraints [32]. Next, the aspects of the BHO for the solution nonlinear programming problems by presenting its main characteristics.

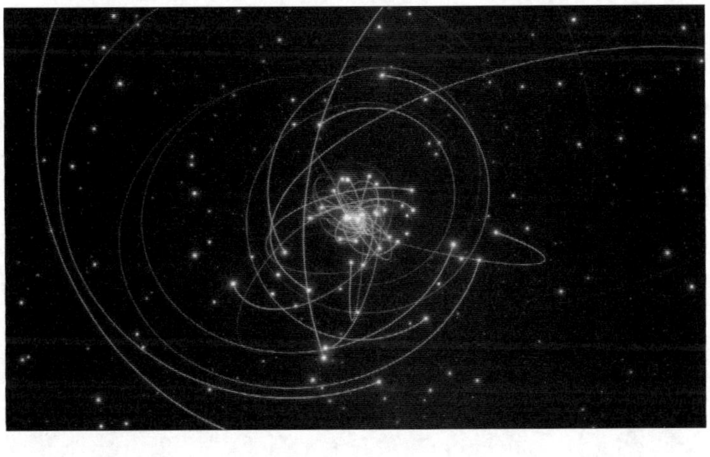

Figure 1. Interaction between starts and black holes (taken from [33])

3.1. Creation of the stars

The BHO is an optimization algorithm that works with a population that evolve through the solution space which have a similar behavior of the classical particle swarm optimization approach [25, 34]. To create the initial population (distribution of the stars about the solution space) is made a random procedure with the following structure:

$$P^t = \begin{bmatrix} p_{11} & p_{12} & \cdots & p_{1n_v} \\ p_{21} & p_{22} & \cdots & p_{2n_v} \\ \vdots & \vdots & \ddots & \vdots \\ p_{n_i1} & p_{n_i2} & \cdots & p_{n_in_v} \end{bmatrix} \tag{7}$$

Where n_i and n_v correspond to the scalars of the the number of individuals and the number of variables. Note that each row of the population is a star positioned on the solution space, where its coordinates are calculated as follows:

$$p_{ij} = p_j^{\min} + \alpha \left(p_j^{\max} - p_j^{\min}\right) \tag{8}$$

Where α is a random number between 0 and 1 generated with a normal distribution, p_{ij} is the active power component of the start i associated with the node j, being $p_j^{\min}$ and $p_j^{\max}$ the minimum and maximum power injections admissible for the DG connected at node j as was defined in Equation (4).

P^t represents the current population, each individual (star) inside of it represents the total generation at all nodes that have connected DG. In addition, the best solution contained in P^t is chosen as the current hole location [25].

3.2. Stars' movements

The movement of each individual, i.e., each star P_i^{t+1}, through the solution star is governed by the current position of this and the current position (p_i^{BH}) of the black hole. The position of the star is modified by applying the Equation (9).

$$P_i^{t+1} = p_i^t + \beta \left(p_i^{BH} - p_i^t\right), \; \forall i = 1, 2, ..., n_i \tag{9}$$

Where β is a random number between 0 and 1 generated with a normal distribution. It is important to highlight that for each component of the new star P_i^{t+1} is mandatory to revise that the upper and lower bounds in Equation (4) must be satisfied; otherwise, Expression (8) is used to guarantee the feasibility of the solution space.

3.3. Updating the position of the black hole

The position of the black hole corresponds to the best solution found in the current population during the search process through the solution space. The position of the black hole will be updated if in the descending population of stars (see Equation (9), the objective function the ith individual is lower than the current black hole, i.e., $z_f\left(P_i^{t+1}\right) < z_f\left(P_i^{BH}\right)$, which implies that $P_{t+1}^{BH} = P_i^{t+1}$. Otherwise, the position of the black hole remains unaltered.

Note that the evaluation of the objective function $z_f(\cdot)$ requires the implementation of a power flow methodology to determine the voltages in the nodes, which allow determining the level of power losses and the voltage regulation bounds. Here is adopted the power flow methodology named the triangular-based power flow method recently reported in [35].

3.4. Stars' substitution

One of the main aspects of the BHO approach is the methodology to replace some of the stars in the current population that no meet the diversity criterion [25]. In the theoretical physics concerning black holes, an stars that is in the neighborhood of the event horizon is destined to the destruction, since this star can be absorbed/or ejected due to the strong gravitational forced acting on it [32]. To represent the possibility that an arbitrary star in P_i^{t+1} is absorbed/ejected by the black hole, the radius of the event horizon can be defined as presented below:

$$EH_R = \frac{\sum_{i=1}^{n_i} z_f\left(P_i^{t+1}\right)}{z_f\left(P_{BH}^{t+1}\right)}, \qquad (10)$$

Where the denominator in Equation (10) represents the best objective function int he current population, i.e., the objective function value of the current black hole, and the numerator in Equation (10) corresponds to the algebraic sum of the objective function associated with all the stars in the population.

To know if an arbitrary star crosses the event horizon, the norm-2 (i.e., Euclidean norm) of such star with respect to the black hole's location is used:

$$D_{BH,i} = \left|\left|P_{BH}^{t+1} - P_i^{t+1}\right|\right|, \qquad (11)$$

Note that if $EH_R \leq D_{BH,i}$, a new star is generated randomly to replace the star absorbed by the black hole; otherwise, the star P_i^{t+1} continues being part of the current population. It is worth mentioning that the birth of new stars increases the possibility of the BHO to explore new regions of the solution space, being the generation of new stars a global exploration criteria [25].

3.5. Stop criterion

The exploration and exploitation ends if one of the following criteria meets:

- ✓ The maximum number of iteration, i.e., $t_{\max}$ is reached.
- ✓ After k consecutive iterations the location of the BH has nor presented any improvement.

4. Test feeders' information

To evaluate the proposed strategy in the problem optimal dimensioning of DG, two classical radial distribution test feeders composed of 33 and 69 nodes are used. Electrical configurations of these test feeders are depicted in Figs. 2(a) and 2(b). The complete information regarding lines, branches and loads for both test feeders can consultedin in [11].

5. Numerical validation

To evaluate the efficiency of the proposed strategy, we evaluate different reports presented in the literature for both test feeders. The methods used for the comparative analysis with the following:

- ✓ Krill-herd Algorithm (KHA) [13].
- ✓ Loss-sensitivity Factor with Simulated Annealing (LSFSA) [36].
- ✓ Genetic alAorithms (GA) [16].
- ✓ Particle Swarm Optimization (PSO) [16].
- ✓ Hybrid GA/PSO [37].

5.1. IEEE 33-bus system

The results of the comparative methods and the black-hole optimization approach applied to the IEEE 33-bus system is described in Table 1. The proposed BHO in all the comparative cases has positive improvements, being the GA/PSO method the worst approach to determine the size of the dispersed generators with a difference 29.11 % when compared with the BHO; while

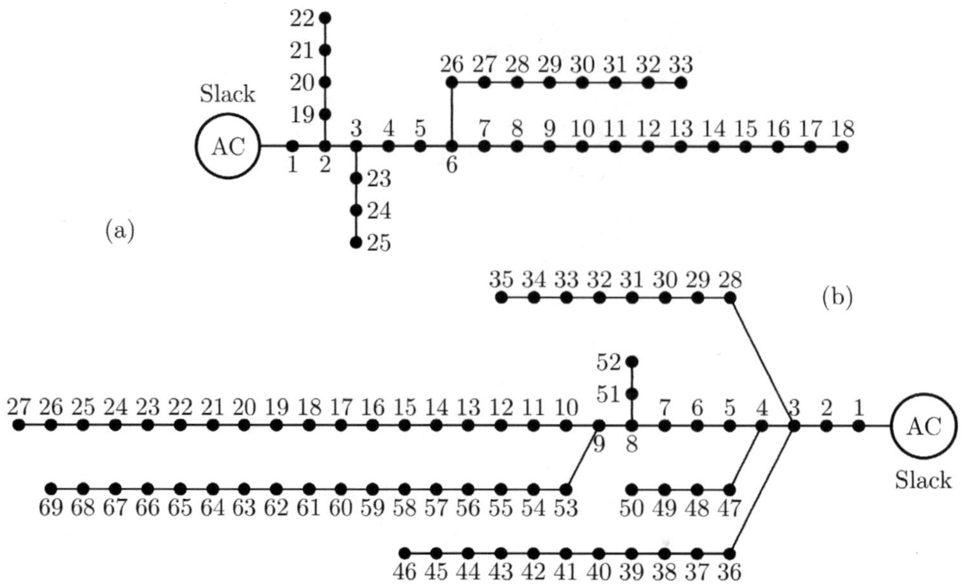

Figure 2. Electrical configurations of the test systems: (a) IEEE 33-bus, and (b) IEEE 69-bus.

Table 1. BHO performance in the IEEE 33-bus system.

Literature reports				Black-hole optimizer		
Method	**Location**	**Size** (pu)	**Losses** (kW)	**Size** (pu)	**Losses** (kW)	**Improv.** (%)
	13	0.8107		0.8129		
KHA	25	0.8368	75.3977	0.8708	73.5036	2.51
	30	0.8410		1.0736		
	6	1.1124		1.2670		
LSFSA	18	0.4874	82.0418	0.4840	81.8575	0.22
	30	0.8679		0.7805		
	11	1.5000		0.9681		
GA	29	0.4228	106.0682	0.5571	86.5173	18.43
	30	1.0714		0.6068		
	8	1.1768		0.7475		
PSO	13	0.9816	105.3296	0.5644	82.9698	21.23
	32	0.8297		0.8388		
	11	0.9250		0.6721		
GA/PSO	16	0.8630	121.3259	0.3738	86.0127	29.11
	32	1.2000		0.9231		

the LSFSA approach only presents a difference about 0.22 % being the second best approach to determine the size of the generators behind of the proposed BHO.

Comparative results in Table 1 show that the selection of the location points for the disperse generators influence directly the possibility of reducing the amount of power losses with respect to the base case; being nodes 11, 29, and 30 with generators with nominal rates of 0.9681 MW, 0.5571 MW, and 0.6068 MW that produce the higher level of power losses with a value of 86.5173 kW, which represent a reduction about 58.99 % with respect the base case, i.e., 210.9876 kW without insertion of dispersed generation. On the other hand, when the disperse generators are located in nodes 13, 25, and 30 with nominal rates of 0.8129 MW, 0.8708 MW, and 1.0736 MW, which produce a total power losses of 73.5036 kW when is used the proposed BHO. This represents a reduction in the amount of power losses with respect to the benchmark case about 65.16 %.

5.2. IEEE 69-bus system

The results of the comparative methods and the black-hole optimization approach for the IEEE 69-bus system is described in Table 2. For this system, it is observed that the BHO presents improvements regarding the final value of the power losses when compared with the literature reports. In this system the second best method corresponds to the KHA with a difference of 0.19 % with respect to the BHO; while the worst method corresponds to the PSO approach with a difference of 18.23 %.

Table 2. BHO performance in the IEEE 69-bus system.

Literature reports				**Black-hole optimizer**		
Method	**Location**	**Size** (pu)	**Losses** (kW)	**Size** (pu)	**Losses** (kW)	**Improv.** (%)
	11	0.4962		0.5486		
KHA	22	0.3113	69.7144	0.3506	69.5841	0.19
	61	1.7354		1.7194		
	18	0.4204		0.5252		
LSFSA	60	1.3311	77.2461	1.3590	76.6236	0.81
	65	0.4298		0.4563		
	21	0.9297		0.4903		
GA	62	1.0752	89.0286	1.4710	73.1102	17.88
	64	0.9925		0.2925		
	17	0.9925		0.5267		
PSO	61	1.1998	83.9634	1.4614	71.6566	18.23
	63	0.7956		0.3210		
	21	0.9105		0.4894		
GA/PSO	61	1.1926	84.7498	1.4701	72.1063	14.92
	63	0.8849		0.3143		

In addition, the effect of the points where the dispersed sources are located have important incidences in the final power losses. For example, for nodes 11, 22, and 61, the BHO reports generation sizes of 0.5486 MW, 0.3506 MW, and 1.7194 MW. These power injections allows a final power losses of 69.5841 kW; i.e., 69.08 % respect to the benchmark case (225.0718 kW). When the distributed sources are assigned at nodes 18, 60, and 65 with nominal rates of 0.5252 MW, 1.3590 MW, and 0.4563 MW, the power losses is reduced to 65.96 %.

6. Conclusions

Was addressed in this research he optimal injection of active power in dispersed generator with black hole optimizer.. This is an easily implementable metaheurisitic optimization technique with few parameters to tune. Numerical results demonstrate that the proposed BHO improves all the final power losses for all the comparative methods in both test feeders. In the case of the IEEE 33-node test system the BHO reaches improvements between 0.22 % and 29.11 % when compared with the LSFSA and the GA/PSO approaches. For the IEEE 69-bus system the proposed approach reaches improvements between 0.19 % and 18.23 % when compared with the KHA and the PSO methodologies.

The incidence of the points where the dispersed sources are located was evidenced in the total reduction of the power losses. For the IEEE 33-bus system, the minimum reduction of the power losses was about 58.99 % with the generators located at nodes 11, 29, and 30; while the best reduction was about 65.16 where the dispersed sources where connected in nodes 13, 25 and 30. For the IEEE 69-bus system when the disperse sources were located in nodes 11, 22, and 61 the reduction of the power losses was about 69.08 % which was the best reduction for this system; while when the distributed sources are located in nodes 18, 60, and 65, the reduction of the power losses was about 65.96 %, i.e., the worst reduction scenario.

References

[1] S.A. ChithraDevi, L. Lakshminarasimman, and R. Balamurugan. Stud krill herd algorithm for multiple DG placement and sizing in a radial distribution system. *Engineering Science and Technology, an International Journal*, 20(2):748–759, apr 2017. doi: 10.1016/j.jestch.2016.11.009.

[2] Muhammad Usama, Mahmoud Moghavvemi, Hazlie Mokhlis, Nurulafiqah Nadzirah Mansor, Haroon Farooq, and Alireza Pourdaryaei. Optimal Protection Coordination Scheme for Radial Distribution Network Considering ON/OFF-Grid. *IEEE Access*, 9: 34921–34937, 2021. doi: 10.1109/access.2020.3048940.

[3] Krischonme Bhumkittipich and Weerachai Phuangpornpitak. Optimal Placement and Sizing of Distributed Generation for Power Loss Reduction Using Particle Swarm Optimization. *Energy Procedia*, 34:307–317, 2013. doi: 10.1016/j.egypro.2013.06.759.

[4] Susanta Dutta, Provas Kumar Roy, and Debashis Nandi. Optimal location of STATCOM using chemical reaction optimization for reactive power dispatch problem. *Ain Shams Engineering Journal*, 7(1):233–247, mar 2016. doi: 10.1016/j.asej.2015.04.013.

[5] Alejandro Valencia, Ricardo A. Hincapie, and Ramón A. Gallego. Optimal location, selection, and operation of battery energy storage systems and renewable distributed generation in medium–low voltage distribution networks. *Journal of Energy Storage*, 34: 102158, feb 2021. doi: 10.1016/j.est.2020.102158.

[6] Joshua A. Taylor and Franz S. Hover. Convex Models of Distribution System Reconfiguration. *IEEE Transactions on Power Systems*, 27(3):1407–1413, aug 2012. doi: 10.1109/tpwrs.2012.2184307.

[7] Sandeep Kaur, Ganesh Kumbhar, and Jaydev Sharma. A MINLP technique for optimal placement of multiple DG units in distribution systems. *Int. J. Electr. Power Energy Syst.*, 63:609–617, dec 2014. doi: 10.1016/j.ijepes.2014.06.023.

[8] Luis Grisales-Noreña, Daniel Gonzalez Montoya, and Carlos Ramos-Paja. Optimal sizing and location of distributed generators based on PBIL and PSO techniques. *Energies*, 11 (4):1018, apr 2018. doi: 10.3390/en11041018.

[9] Walter Gil-González, Alejandro Garces, Oscar Danilo Montoya, and Jesus C. Hernández. A Mixed-Integer Convex Model for the Optimal Placement and Sizing of Distributed Generators in Power Distribution Networks. *Applied Sciences*, 11(2):627, jan 2021. doi: 10.3390/app11020627.

[10] Oscar Danilo Montoya, Alexander Molina-Cabrera, Harold R. Chamorro, Lazaro Alvarado-Barrios, and Edwin Rivas-Trujillo. A Hybrid Approach Based on SOCP and the Discrete Version of the SCA for Optimal Placement and Sizing DGs in AC Distribution Networks. *Electronics*, 10(1):26, dec 2020. doi: 10.3390/electronics10010026.

[11] Oscar Danilo Montoya, Walter Gil-González, and L.F. Grisales-Noreña. An exact MINLP model for optimal location and sizing of DGs in distribution networks: A general

algebraic modeling system approach. *Ain Shams Eng. J.*, 11(2):409–418, jun 2020. doi: 10.1016/j.asej.2019.08.011.

[12] Walter Gil-González, Oscar Danilo Montoya, Luis Fernando Grisales-Noreña, Alberto-Jesus Perea-Moreno, and Quetzalcoatl Hernandez-Escobedo. Optimal Placement and Sizing of Wind Generators in AC Grids Considering Reactive Power Capability and Wind Speed Curves. *Sustainability*, 12(7):2983, apr 2020. doi: 10.3390/su12072983.

[13] Sneha Sultana and Provas Kumar Roy. Krill herd algorithm for optimal location of distributed generator in radial distribution system. *Applied Soft Computing*, 40:391–404, mar 2016. doi: 10.1016/j.asoc.2015.11.036.

[14] Walter Gil-González, Oscar Danilo Montoya, Luis Fernando Grisales-Noreña, Carlos Alberto Ramírez Vanegas, and Alexander Molina Cabrera. Hybrid Optimization Strategy for Optimal Location and Sizing of DG in Distribution Networks. *Tecnura*, 24(66):47–61, oct 2020. doi: 10.14483/22487638.16606.

[15] D. Sattianadan, M. Sudhakaran, S. S. Dash, K. Vijayakumar, and P. Ravindran. Optimal Placement of DG in Distribution System Using Genetic Algorithm. In *Swarm, Evolutionary, and Memetic Computing*, pages 639–647. Springer International Publishing, 2013. doi: 10.1007/978-3-319-03756-1_57.

[16] M.H. Moradi and M. Abedini. A combination of genetic algorithm and particle swarm optimization for optimal DG location and sizing in distribution systems. *International Journal of Electrical Power & Energy Systems*, 34(1):66–74, jan 2012. doi: 10.1016/j.ijepes.2011.08.023.

[17] Oscar Danilo Montoya, Walter Gil-González, and César Orozco-Henao. Vortex search and Chu-Beasley genetic algorithms for optimal location and sizing of distributed generators in distribution networks: A novel hybrid approach. *Engineering Science and Technology, an International Journal*, 23(6):1351–1363, dec 2020. doi: 10.1016/j.jestch.2020.08.002.

[18] Mohamed A. Tolba, Vladimir N. Tulsky, and Ahmed A. Zaki Diab. Optimal allocation and sizing of multiple distributed generators in distribution networks using a novel hybrid particle swarm optimization algorithm. In *2017 IEEE Conference of Russian Young Researchers in Electrical and Electronic Engineering (EIConRus)*. IEEE, 2017. doi: 10.1109/eiconrus.2017.7910880.

[19] R V S Laksmi Kumari, G V Nagesh Kumar, S Siva Nagaraju, and M. Babita Jain. Optimal sizing of distributed generation using particle swarm optimization. In *2017 International Conference on Intelligent Computing, Instrumentation and Control Technologies (ICICICT)*. IEEE, jul 2017. doi: 10.1109/icicict1.2017.8342613.

[20] Thuan Thanh Nguyen. Enhanced sunflower optimization for placement distributed generation in distribution system. *International Journal of Electrical and Computer Engineering (IJECE)*, 11(1):107, feb 2021. doi: 10.11591/ijece.v11i1.pp107-113.

[21] K. Nara, Y. Hayashi, K. Ikeda, and T. Ashizawa. Application of tabu search to optimal placement of distributed generators. In *2001 IEEE Power Engineering Society Winter Meeting. Conference Proceedings (Cat. No.01CH37194)*. IEEE, 2001. doi: 10.1109/pesw.2001.916995.

[22] M. Gandomkar, M. Vakilian, and M. Ehsan. A genetic–based tabu search algorithm for optimal DG allocation in distribution networks. *Electric Power Components and Systems*, 33(12):1351–1362, dec 2005. doi: 10.1080/15325000590964254.

[23] Eyad S. Oda, Abdelazeem A. Abdelsalam, Mohamed N. Abdel-Wahab, and Magdi M. El-Saadawi. Distributed generations planning using flower pollination algorithm for enhancing distribution system voltage stability. *Ain Shams Eng. J.*, 8(4):593–603, dec 2017. doi: 10.1016/j.asej.2015.12.001.

[24] Suresh Kumar Sudabattula and M. Kowsalya. Flower Pollination Algorithm Based Optimal Placement of Solar Based Distributed Generators in Distribution System . *International Journal of Renewable Energy Research-IJRER*, 6(4):1–10, June 2016. URL `https://www.ijrer.org/ijrer/index.php/ijrer/article/view/4364`.

[25] Orfilio Sebastian Velasquez, Oscar Danilo Montoya, Victor Manuel Garrido, and Luis Fernando Grisales. Optimal Power Flow in Direct-Current Power Grids via Black Hole Optimization. *Advances in Electrical and Electronic Engineering*, 17(1), mar 2019. doi: 10.15598/aeee.v17i1.3069.

[26] Abdolreza Hatamlou. Black hole: A new heuristic optimization approach for data clustering. *Information Sciences*, 222:175–184, feb 2013. doi: 10.1016/j.ins.2012.08.023.

[27] Ricardo Soto, Broderick Crawford, Rodrigo Olivares, Carla Taramasco, Ignacio Figueroa, Álvaro Gómez, Carlos Castro, and Fernando Paredes. Adaptive Black Hole Algorithm for Solving the Set Covering Problem. *Mathematical Problems in Engineering*, 2018:1–23, oct 2018. doi: 10.1155/2018/2183214.

[28] Masoum Farahmandian and Abdolreza Hatamlou. Solving optimization problems using black hole algorithm. *Journal of Advanced Computer Science & Technology*, 4(1):68, feb 2015. doi: 10.14419/jacst.v4i1.4094.

[29] Junqi Zhang, Kun Liu, Ying Tan, and Xingui He. Random black hole particle swarm optimization and its application. In *2008 International Conference on Neural Networks and Signal Processing*. IEEE, jun 2008. doi: 10.1109/icnnsp.2008.4590372.

[30] R. Balamurugan, A. M. Natarajan, and K. Premalatha. Stellar-mass black hole optimization for biclustering microarray gene expression data. *Applied Artificial Intelligence*, 29(4):353–381, apr 2015. doi: 10.1080/08839514.2015.1016391.

[31] Luis Fernando Grisales-Noreña, Oscar Daniel Garzon-Rivera, Oscar Danilo Montoya, and Carlos Andrés Ramos-Paja. Hybrid Metaheuristic Optimization Methods for Optimal Location and Sizing DGs in DC Networks. In *Communications in Computer and Information Science*, pages 214–225. Springer International Publishing, 2019. doi: 10.1007/978-3-030-31019-6_19.

[32] H.R.E.H. Bouchekara. Optimal power flow using black-hole-based optimization approach. *Appl Soft Comput*, 24:879–888, nov 2014. doi: 10.1016/j.asoc.2014.08.056.

[33] eoPortal. Black hole and its shadow - first direct visual evidence of a supermassive black hole. Online, October 2020. URL `https://directory.eoportal.org/web/eoportal/satellite-missions/b/black-hole`.

[34] Adam P. Piotrowski, Jaroslaw J. Napiorkowski, and Pawel M. Rowinski. How novel is the "novel" black hole optimization approach? *Information Sciences*, 267:191–200, may 2014. doi: 10.1016/j.ins.2014.01.026.

[35] Abbas Marini, S.S. Mortazavi, Luigi Piegari, and Mohammad-S. Ghazizadeh. An efficient graph-based power flow algorithm for electrical distribution systems with a comprehensive modeling of distributed generations. *Electric Power Systems Research*, 170:229–243, may 2019. doi: 10.1016/j.epsr.2018.12.026.

[36] Satish Kumar Injeti and N. Prema Kumar. A novel approach to identify optimal access point and capacity of multiple DGs in a small, medium and large scale radial distribution systems. *International Journal of Electrical Power & Energy Systems*, 45(1):142–151, feb 2013. doi: 10.1016/j.ijepes.2012.08.043.

[37] Oscar Danilo Montoya, Luis Fernando Grisales-Noreña, William Tadeo Amin, Luis Alejandro Rojas, and Javier Campillo. Vortex Search Algorithm for Optimal Sizing of Distributed Generators in AC Distribution Networks with Radial Topology. In *Communications in*

Computer and Information Science, pages 235–249. Springer International Publishing, 2019. doi: 10.1007/978-3-030-31019-6_21.

AUTHOR INDEX